Build Your Own Intelligent Amateur Radio Transceiver

Build Your Own Intelligent Amateur Radio Transceiver

Randy Lee Henderson

McGraw-Hill

New York   San Francisco   Washington, D.C.   Auckland   Bogotá   Caracas   Lisbon   London
Madrid   Mexico City   Milan   Montreal   New Delhi   San Juan   Singapore   Sydney   Tokyo   Toronto

Library of Congress Cataloging-in-Publication Data

Henderson, Randolph L.
    Build your own intelligent amateur radio transceiver / Randy Lee
Henderson.
        p.   cm.
    Includes index.
    ISBN 0-07-028263-3 (h). — ISBN 0-07-028264-1 (p)
    1. Radio—Transmitter-receivers—Design and construction—
Amateurs' manuals.   2. Shortwave radio—Receivers and reception—
Design and construction—Amateurs' manuals.   I. Title.
TK9956.H397   1997
621.384'131—dc20   96-41801
                                                            CIP

McGraw-Hill

A Division of The McGraw·Hill Companies

1 2 3 4 5 6 7 8 9 0   DOC/DOC   9 0 1 0 9 8 7 6

ISBN 0-07-028263-3 (HC)

ISBN 0-07-028264-1 (PBK)

*The sponsoring editor for this book was Steve Chapman, the editing supervisor was
Scott Amerman, and the production supervisor was Don Schmidt. It was set in
ITC Century Light by North Market Street Graphics.*

Printed and bound by R. R. Donnelley & Sons Company.

McGraw-Hill books are available at special quantity discounts to use as premiums and
sales promotions, or for use in corporate training programs. For more information,
please write to the Director of Special Sales, McGraw-Hill, 11 West 19th Street,
New York, NY 10011. Or contact your local bookstore.

To my mother and father

Contents

Appendices

Preface

THIS BOOK IS ABOUT BUILDING HIGH-FREQUENCY (HF) amateur radio gear and test equipment. I hope it will enable you to build, enjoy, and understand a sophisticated radio transceiver without being overwhelmed by the many details involved in its development. It's about radios you would want to use, not curios to be dusted off and admired occasionally.

Why build HF gear? It's certainly not of necessity. That has long ago ceased to be a valid reason. Although the transceiver described here may cost less to build than new entry-level radios, it can cost as much as older, high-quality used rigs.

Obviously, the motivation for building and experimenting lies elsewhere. Curiosity and the quest for knowledge is certainly high on the list. There's always the challenge to do things better, or more cheaply, easily, or attractively. Some projects have custom features not available in mass-produced equipment. Just imagine—you have access to "factory-authorized" repair and modification services at a moment's notice. Operating equipment born of your own creative efforts can be very satisfying.

The course of entry into amateur radio has changed. The initiation ritual of pounding brass on the novice continuous-wave (CW) bands is not as common now. That's too bad for beginning experimenters, because other modes often require more complex construction. Don't get the wrong idea—the first paragraph is in no way a condemnation of simple CW transmitters and receivers. They're great. Every ham should try them. They can give a tremendous return in performance for the time and money invested. It's a fine place to start.

The material contained in the chapters that follow is a result of my attempt to go beyond this starting point, while using commonly available parts and techniques. Some of the operational capabilities once categorized as "bells and whistles" are now almost considered necessities. A necessity might be thought of as something that makes the difference between QSOs (contacts) and disappointments.

Propagation often necessitates multiband capability. Scheduled contacts and nets may require an accurate frequency display. Some DX contacts are impossible without split-frequency operation. The frequency agility of memories or registers allows you to catch that rare station before it's too late.

Much of this book is arranged as a series of small articles on construction. The end goal is to use them to assemble a multiband, multimode, microprocessor-controlled HF transceiver. If that sounds like a quest worthy of experienced experimenters, I suppose it is. However, printed circuit patterns reduce the risk of error.

A neophyte builder can succeed by enlisting the aid of a competent mentor. If you prefer an appetizer to the main course, interesting radios can be created from a fraction of the projects. This book uses a building-block approach that starts at the speaker and microphone, and heads toward the antenna. Examples of useful test equipment are included.

I will assume the reader is familiar with at least the most basic electrical fundamentals and electronic components (or is willing to acquire this knowledge). A narrative of signal flow is given in the appropriate places. However, more instructional material is devoted to frequency synthesis and microcomputer/microcontroller subjects.

Building in a lot of features is challenging. Putting many capabilities into a reasonably priced piece of gear always involves compromises (regardless of what advertisers say). For instance, a number of synthesizer designs have been published, but some have outputs so dirty you wouldn't want them in your garbage can, much less your receiver.

Because so much current gear uses synthesized methods of frequency control, I think it is a good subject to examine more deeply. Various methods of frequency synthesis are illustrated. Methods of interfacing between the operator, microprocessor, and radio circuits are discussed. I hope you find the examples useful in adding features to your home-built gear.

When I hear hams talk about building, one lament seems to surface regularly: "You just can't find parts any more." Well, that could be true if you're looking for obsolete parts or expect to obtain them all from one supplier. As with everything else, technology changes. Things are done differently now.

Commercial fabricators have resources not available to the average home experimenter. That doesn't mean you can't do as well or better; it just means you will have to do it differently.

I suspect most who build have learned the aforesaid and find replacements for many parts. Parts described herein are relatively common and inexpensive. They are available from multiple vendors and, in most cases, from multiple manufacturers.

Most of the book is a guide to building a modern microprocessor-controlled ham transceiver. By actually building this radio, I've faced and conquered a number of knotty problems encountered by people trying to build their own gear. The book not only describes construction techniques, it explains the operation and purpose of each circuit.

There are books and magazine articles that describe the theory and design principles involved in modern radio transceivers, leaving the reader to work out the details. Other sources give concrete data for building the functional blocks of high-tech circuits, such as frequency synthesizers, but not entire radios. Hobbyists can even find publications describing complete transceiver construction projects that use obsolete methods and technology. The situation mentioned above is why I've written this book.

In the past, amateurs built equipment that often equaled or exceeded the features found in commercial equipment. Current economic incentives for building are less dominant; however, the desire people have to master technology is the same or greater. As far as I can tell, there is no integrated source of practical information that will allow hobbyists to build radio equipment with the features used by most hams and shortwave listeners today. This book will help fill that gap.

The entire transceiver circuit is described visually and in text. Firmware information is included that will allow you to understand the power you can wield by controlling your projects this way. Advanced hobbyists can take advantage of all the legwork I've done. You can make completely new projects by modifying firmware and/or hardware. The book also has some added benefits such as smaller projects if you don't want to tackle a multiband, multimode HF transceiver.

Chapter 1 covers all of the audio frequency circuits. The transmit/receive switching scheme is explained. Signal flow is traced through the amplifiers, and the microphone is discussed, along with information about the audio frequency response of single-sideband (SSB) communication circuits. Information about circuit-board construction and testing completes the chapter.

Chapter 2 deals with circuits that, when combined with audio circuits, produce double-sideband voice and continuous-wave telegraphy signals. Topics include mixers, oscillators, and transformer winding. This information is expanded to implement simple receivers and transceivers. Bandspread tuning for the simple receiver and transceiver circuits is discussed.

Chapter 3 concerns lower-frequency IF circuits. Transmit- and receive-mode signals are described. A peak clipper or limiter is explained. I include information about homemade crystal filters and construction tips. The chapter finishes with an audio generator construction project.

Chapter 4 covers the high-intermediate-frequency circuits and includes discussions of discrete transistor bidirectional amplifiers, more on mixers, an LC (inductance/capacitance) filter, a crystal filter, and the second local oscillator.

Chapter 5 looks at circuits that, for the majority, operate directly at the radio frequency of interest. The sections include the power amplifier, driver, T/R switches, and band-switched filters. A control/interface board is included.

Chapter 6 is about how and why we generate high-frequency signals. Topics include early methods, coherent sources, modulation, and direct digital synthesis.

Chapter 7 provides some practical examples of transistor oscillator circuits. To facilitate comparisons, the oscillator circuits are all designed for a similar frequency range. The advantages of different circuits are discussed, and practical construction considerations are pointed out.

Chapter 8 is a basic intuitive introduction to the phase-locked loop (PLL) circuit. Simple logic elements using bipolar transistors serve as building blocks to explain how to make the flip-flops and, ultimately, the programmable counters used in a PLL synthesizer.

Chapter 9 describes how prescalers are used for extending the upper frequency range of a PLL. They are discussed along with dual-modulus prescaling. A simplified form of frequency synthesizer using an alternative method is explained.

Much of the actual frequency synthesizer used in the transceiver project is dealt with in chapter 10. This chapter shows how a

xvi

digital-to-analog converter (DAC) controls the variable-frequency reference oscillator and how hardware in the microcontroller controls the fine-tune loop.

A novel voltage-controlled oscillator and control circuits complete the coarse-tune loop of the synthesizer in chapter 11. Mechanical details are included.

In chapter 12, facts about binary numbers help form a background for discussing the microcontroller. Construction information for the computer board is included.

Chapter 13 starts with constructing the optical shaft encoder. This mechanism connected to the main tuning knob is described, as is the software decoding scheme for reading position and travel of the rotary shaft. Along with the necessary wiring information, layout and operation of the front panel switches is explained. Detailed information for communicating with the liquid-crystal display module is included.

Chapter 14 begins with some very elementary ideas about programming. A condensed list of the instruction set for the microcontroller is included.

Chapter 15 describes what the microcontroller firmware must do and how some of the more interesting tasks are accomplished. It also includes information useful for readers who want to modify and develop firmware.

In chapter 16, potential problems related to mechanical design are examined. The text explains why the enclosure design is important and gives helpful tips. The rest of the chapter is full of ideas for modifying or expanding the transceiver project.

Chapter 17 has information about how to build a rudimentary spectrum analyzer attachment for oscilloscopes. It includes some construction details and ideas about how it could be expanded and improved.

Chapter 18 first appeared as a magazine article and is used here with a bit of additional material. It describes the hardware used to test and evaluate crystal filters used in the transceiver project.

Chapter 19 first provides information and some generalized advice about operations using CW and SSB modes. Later, issues specific to operating the project transceiver are covered.

I have searched for information that would allow me to do some of the things the big boys do when it comes to building radio gear.

Unfortunately, it just isn't practical to duplicate some factory processes at home—at least not at this home. I found alternatives that don't require proprietary parts or processes.

The quartz crystal filters were one concern. Commercial filters, although expensive, are available in small quantities. I provide options for builders who want to use them. However, by using the homemade alignment gear, excellent filters can be fabricated from inexpensive microprocessor-clock crystals.

The synthesizer was a more difficult challenge. A conventional four-loop synthesizer is a large, messy, complicated affair even without the microprocessor that controls it. The configuration I chose actually uses timers built into the 80C31 microcontroller to accurately measure the period of a signal in the second (fine-tune) loop. An interrupt-driven software routine calculates the frequency. This information is manipulated in order to drive a digital-to-analog converter that corrects the frequency of a voltage-controlled crystal oscillator.

This crystal-controlled VCO is the reference-frequency signal for the coarse-tune loop. The performance of this scheme is remarkable. Even though the fine-tune loop can resolve frequency steps of less than 1 Hz, it achieves frequency lock in a fraction of a second.

Of course, a direct digital synthesis (DDS) reference is faster, but it would be more expensive and, in some respects, more complicated. Operators who have used the rig reported they liked the feel and response of its tuning.

Most of the book is directed at getting things done, not proposing what could be done. For this reason, construction starts at the microphone/speaker end and works toward the antenna. Doing so allows you to test and debug a section at a time. Previously completed stages help provide the signals and other resources needed to test succeeding stages.

In order to use the information in this book, you should have a basic understanding of ac and dc fundamentals. You should also know how to use common active devices such as bipolar transistors. The ability to use basic tools and test equipment is necessary. In other words, you should have some previous experience constructing electronics projects.

Students and instructors may find this book useful. Much background material is included to help the reader form a better understanding of frequency synthesis. Although most students may not have time for a large hobby project, this book could be a valuable resource in some courses.

Because the book breaks down various sections of the transceiver into building blocks, it is well suited to group projects in which individuals or small groups complete and test separate sections. Each small group or individual could then cooperate with the class in assembling and testing an entire transceiver. This is similar to real-world situations in which technical and personal communication skills are important in completing a project.

Operating commercially made ham gear is enjoyable (I even have some around here, I think), but it doesn't compare with taking credit for a signal of your own making. When you also hear stations because of your own creative efforts, will you ever be able to hit the off switch? Let's find out.

Acknowledgments

I WISH TO THANK NUMEROUS AMATEURS WHO HELPED make this project a reality.

My wife Sylvia deserves special thanks. In addition to providing moral support and encouragement, she is a licensed amateur and has used a number of my homebrew projects on the air. Although she is not a builder, her curiosity and willingness to operate some unusual gear have allowed me to observe and evaluate projects more effectively than when I use them by myself. It's also more satisfying to share the fruits of the labor (and more exasperating when one of my building mistakes makes itself apparent in the middle of one of her QSOs).

John Carter, W5LGO, has been very helpful in a number of ways. He has provided advice on various topics and assistance with testing some of the gear in the book. Thanks also go to Dub Thornton, WA5YFY, who has helped with some computer-related advice, and James "Bunky" Botts, K4EJQ, who helped me get started in amateur radio.

xxi

Audio circuits

THE MICROPHONE AND SPEAKER SEEM A LOGICAL PLACE to start for a number of reasons. For communication modes other than digital data transfer, these are the beginning and ending points for signal flow. The audio circuits are a bit easier to understand than some other areas of a transceiver. Many sections of transceivers set forth in this book are bidirectional, with transmitted signals going in one direction and received signals doing the opposite. It's also the starting point for a simple phone (double-sideband, or DSB) transceiver or a more sophisticated rig.

The transmit and receive scheme of using a two-wire control line or bus is shown in figure 1-1. The control line is simply a wire that supplies power (positive 12 V) or a ground, depending on the requirements of the circuit it controls. Most circuits controlled in the transceiver are activated with a high (positive voltage) and deactivated by a low (ground potential). Some of the low-power amplifier circuits are powered directly by the control line. High power stages and other circuits use the line as a bias supply.

Figure 1-2 displays the audio amplifier schematic for both transmit and receive functions, along with the associated switching circuits. Transmit or receive audio is available at C1. The negative end of C1 is connected to the product detector/balanced modulator. C1 is one of several points in the transceiver where the signal flow for received or transmitted signals takes exactly the same route. This is possible with some circuits such as the detector/modulator, mixers, and filters. The need for a duplicate set of circuits is eliminated; however, complexity is added by the need to switch and route signals.

Transmit/receive switching

Some of the circuits are expensive or difficult enough that you won't want to duplicate them if a switching scheme can be implemented. Crystal filters are usually expensive if purchased. The single-

TRANSMIT

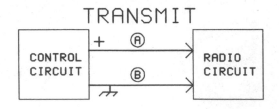

RECEIVE

■ **1-1** *Most of the transceiver sections use two control lines to change between transmit and receive modes. The ground symbol is drawn to indicate that the control circuit provides a low-resistance path to ground when positive voltage is removed.*

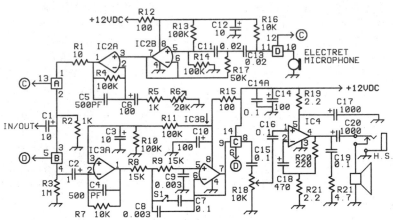

■ **1-2** *Unless otherwise noted, all resistors are 5 percent tolerance; capacitance is in microfarads (μF) and resistance is in ohms (Ω). C14a, C15, C19—0.1-μF capacitors are monolithic-ceramic with 0.1-in lead spacing. The other 0.1-μF capacitors can be any type. Electrolytic-capacitor voltage ratings may range from 16 to 25 Vdc. IC1—4066 quad bilateral switch. IC2 and IC3—TL072 or TL082 opamps. IC4—LM383, TDA2002, or TDA2002H audio power amp. R6—single-turn trimmer potentiometer. Mouser stock number 594-63S203 side adjust or 594-75P-20K top adjust. Switches in IC1 are indicated by a capital letter enclosed by a small square. Pin 7 of IC1 is grounded.*

sideband (SSB) filter discussed in chapter 3 can be time consuming to align. Diode mixers in the transceiver are already bidirectional. These considerations help offset complications introduced by the switching arrangement.

Audio amplifiers can be simply powered up or down to manage the need for transmit or receive operation. Some of the radio-frequency (RF) and intermediate-frequency (IF) amplifiers are controlled in this manner. There is a penalty for doing so with the audio amplifiers. The response time necessary for charging and discharging some of the capacitors is too long. Decoupling networks such as C17 and R19 or C12 and R12 are usually necessary when operating high-gain amplifiers from the same power supply. For this reason, an electronic switch is incorporated in the audio board.

IC1 routes audio in the appropriate direction. The 4066 has been around for a long time and is quite available. It's called a CMOS quad bilateral switch. I will try to explain what that mouthful means. CMOS simply stands for complementary metal-oxide semiconductor. Its insides are made of N-channel and P-channel metal-oxide insulated-gate field-effect transistors. They're arranged into a circuit that forms four independent solid-state switches that allow signals to pass through in either direction.

The control pin turns the switch on and off. For instance, the B switch is controlled by pin 5. When pin 5 has a positive voltage applied to it, the switch is closed; signals pass through. When pin 5 is low or at ground potential, signals are blocked. It's a handy device for switching signals in this and many other circuits.

At this point I should mention something about IC1. In phone-transmit modes, switches A and D are turned on, while switches B and C are turned off. The reverse is true in phone-receive modes. For continuous-wave (CW) telegraphy, switches A and D stay off continuously, while switches B and C stay on to allow the keying side-tone to be heard.

Signal flow for reception

We went through IC1b, so let's proceed for now in the phone-receive mode. IC3a provides the first block of gain. The usual technique for using an op-amp such as IC3a as an audio amplifier with a single-sided supply is to bias the noninverting input (pin 3) to half the supply voltage. Signals arriving at the inverting input (pin 2) are then amplified as the instantaneous voltage at pin 2 varies in relation to pin 3. Because no direct current flows through coupling capacitor C2, the quiescent, or zero point, for audio signals is effectively the same as for pin 3. With the dc potential of pin 3 positioned midway between supply and ground, a separate negative supply is not needed.

Gain in the IC3a stage is affected by the ratio of R7 to the total impedance of components in the input-signal path. By input-signal path, I mean C2, IC1b, C1, and the product detector. IC3b is the active part of a low-pass filter. With S1 open, the cutoff frequency is about 3500 Hz with a flat passband response. Close S1 and the passband peaks around 600 Hz. The 600-Hz peak works well for CW reception and is even useful for knocking down interference from SSB signals that happen to be in the upper audio spectrum.

The received signal must pass through IC1c before proceeding. The reason for this redundant solid-state switch is that the aggregate gain of the audio receive section is higher than the attenuation capability of one switch. R18 is the volume control, the last signal-controlling element before getting to that magnificent power amplifier.

I suppose I'm getting a bit carried away describing the audio output stage, but I really like it. Suggested to me by John Carter, W5LGO, the LM383 is available in 8- and 12-W versions—gobs of power, although that's not why I like it so well. It's the lack of distortion at normal listening levels. In addition to being rugged and easy to mount, it sounds great. Having so much additional peak-power capability seems to be especially helpful on SSB, where the peak-to-average power ratio in an audio signal is much higher than with CW (discounting silent key-up occurrences).

The 1000-µF output capacitor C20 is larger than necessary for most communication needs unless you like listening to really low-pitched CW beat notes. A value of 220 or 100 µF is suitable for voice and normal CW reception.

An equivalent number for this integrated circuit is TDA2002. Both chips include built-in protection against output short circuits and thermal overload.

The microphone

The speech amplifier is designed to use an electret microphone. An electret microphone operates in a manner similar to that of the old condenser microphones used in early radio broadcast studios. Both types of microphones are types of capacitors which change capacitance as sound waves move a flexible diaphragm. The diaphragm forms one side of the capacitor and a fixed plate or housing is used as the other.

Condenser microphones were known for their fidelity but required a high-voltage dc source across the capacitor plates for operation. An electret is an insulator that has a quasi-permanent static electric charge trapped in or on it. The electret acts as the plates would and, being charged, it requires no voltage. The audio signal generated is applied to the gate of a field-effect transistor (FET) housed in the microphone element.

Electret microphones are inexpensive and plentiful, and they have good fidelity. Output levels are higher than with dynamic microphones, allowing lower gain in the speech amplifier. This may lessen the possibility of radio frequency (RF) fields affecting the speech amplifier. The main problem seems to be in responding to low-frequency sounds. Electrets simply do it too well. I'm sure that's great if you are recording music concerts or making nature studies, but all that low-frequency energy robs your signal of power in a sneaky sort of way. Frequencies from 300 to 2500 Hz carry most of the information contained in speech. Sounds below 300 Hz contain a great deal of energy and will use a lot of the available transmitter power if we let them.

You can buy a ready-made microphone for this project or make one. The microphone shown in figure 1-3 continues the theme of building everything yourself. Many electronics suppliers carry electret microphone elements. They are also available in various pieces of salvaged electronics gear.

Almost any arrangement for mounting the element will give usable results, but knowledge of a few principles can result in a better homemade microphone. It is best to mount the element in some sort of soft material rather than clamp or glue it to a rigid housing. I used a rubber grommet. Silicone adhesive or urethane foam is another possibility. This will reduce noise pickup caused by objects bumping or touching the housing.

Try to use a mechanically solid housing. Some inexpensive hand-held microphones crunch, squeak, and pop as operators adjust their grip on the microphone. The push-to-talk switch linkage is also another frequent cause of the crunchy microphone syndrome. Find a switch with a quiet, positive action and stay away from complicated switch linkages if you want to make a quiet, simple microphone.

■ **1-3** *Another form of recycling? Yes, the microphone housing is a discarded plastic bottle. The little thick-walled container is 2½ in long with a diameter of 1½ in. The electret element is similar to Radio Shack's stock number 270-090.*

The shape and size of a housing has a pronounced effect on how the microphone element responds. Sound travels about 335 m/s through the air that is likely to be present when you are using the microphone. A 3000-Hz tone travels 335 m per 3000 vibrations or 11.17 cm per vibration. This is important because the air column in a tube or cylinder open at both ends vibrates strongly when excited by sound waves approximately twice the length of the air column. This resonant length is about 5.58 cm for the 3000-Hz example.

The photo in figure 1-3 shows a homemade microphone constructed from a small plastic container and screw-on lid. A push-button switch is mounted on the side. The electret element is mounted in a rubber grommet, which is in turn mounted in a hole drilled into the lid. Seven holes with a diameter of ¼ in are drilled into the bottom of the container, along with two smaller holes for the audio and switch cords. A neater installation will use a single multiconductor cable. Regardless of whether you use one cable or two, use a shielded pair for the audio and ground lead.

The all-salvage microphone shows definite directional characteristics. Pointing the element end at a sound source gives the strongest response, with sensitivity falling off at the sides. The weakest response is at the back of the microphone, as it should be for a noise-canceling mike. Frequency response is flat enough for good-sounding communications audio. Because of the length of the body, there is a dip in response around 2800 Hz. The range of frequencies affected is very narrow and near the upper end of the voice audio spectrum, so it's not objectionable.

Speech amplifier

Electret microphones require a dc voltage source. Power for their internal FET is supplied by R16. Although the speech amplifier is designed to use an electret microphone, other audio sources can be used. If you need to apply tones to the speech amplifier for digital modes, some sources already have a blocking capacitor in the output circuit, or you can add one along with an accessory jack. Amplified versions of dynamic microphones will also work with this speech amplifier.

After passing through another 4066 switch, the microphone signal (ac component) goes to a high-pass filter built around IC2b. Resistors R13 and R14 serve a dual purpose. They bias IC2 output pins to approximately half the supply voltage to allow for positive and negative swings in the audio waveform. They also form part of a high-pass filter network. If you decide to change the filter characteristics, remember to use equal values for R13 and R14.

The high-pass filter built around IC2b has a cutoff frequency of about 160 Hz with the component values shown. You could use a cutoff frequency as high as 300 Hz. With C11 equal to C13 and R17 equal to the resistance constituted by R13 and R14 in parallel, the cutoff frequency is calculated by the following formula.

$$\text{Frequency} = \frac{1}{2\pi RC} \qquad (1\text{-}1)$$

where frequency is in hertz.

Speech components below the cutoff frequency are reduced to more appropriate levels before being amplified by IC2a. Gain for IC2a is determined by the ratio of R4 to the series string R5, R6. If R4 is the feedback resistor R_f and R5 plus R6 make up the resistance of input resistor R_i, voltage gain for a noninverting amplifier such as IC2a is given by:

$$\text{Voltage gain} = 1 + \frac{R_f}{R_i} \qquad (1\text{-}2)$$

In other words, if you want to use a different-value potentiometer for R6 while retaining the same maximum and minimum gain settings, simply change R4 and R5 in the same proportion. For instance, if you want to use a 10-kΩ pot, reduce R4 and R5 by half. IC1a allows the speech signal to reach the balanced modulator/detector.

Yes, C5 and C6 affect the gain also, but not very much in this case. At audio frequencies in the speech spectrum, the reactance of C5 is high enough to be considered an open circuit and you can consider C6 a short (for ac).

While we're at it, how about inverting amplifiers such as IC3a? What is the relationship? In this case, R7 is the feedback resistor R_f. If a simple resistor R_i (to simulate all the stuff mentioned earlier, such as C2 and IC1b) is connected between pin 2 and a signal source, voltage gain is:

$$\text{Voltage gain} = \frac{R_f}{R_i} \tag{1-3}$$

Audio board

Figure 1-4 shows where the parts are located once soldered in place. IC4 should be fastened to a small heat sink. This part looks like a TO220-style package, except it is made with five leads. The heat sink should also be bolted to the board to eliminate mechani-

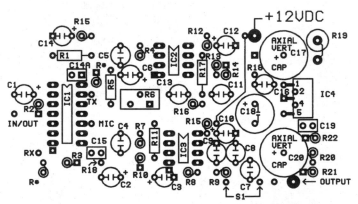

■ **1-4** *The R* 1-MΩ resistors are used for testing purposes when manually switching leads. They have no effect when external switching circuits are installed and may be included or deleted. (actual size.)*

cal stress on the leads caused by shock or vibration. Unless you plan on using lots of audio volume, 2 or 3 in^2 of 0.060-in-thick aluminum should be sufficient heat-sink area.

Transmit audio levels are controlled by R6. A trim pot can be mounted on the board or a panel-mounted pot can be used. If you want to control speech amplifier gain from the front panel, be sure to use shielded wire to connect it. S1, the audio-filter switch, and R18, the volume control, are supposed to be panel-mounted. Use shielded wire for those connections also.

I somehow managed to damage IC1 while testing some other boards. IC sockets are handy if you need to replace a chip.

Figure 1-5 is the resist pattern for the audio board. Double-sided copper-clad stock is used here and in most other boards for the transceiver. Working with the double-sided board at home is easy if you use the right techniques. Most of the boards use one side for circuit paths and the other side is simply a ground plane. A mirror image of the word *audio* means you are looking at the component side of the board in figure 1-5*a*.

Why include a mirror-image pattern? There is a method of transferring patterns to the board by using transparencies made on a plain-paper copy machine (refer to Grebenkemper in the Bibliography). It takes some practice but it skips some of the steps and apparatus involved with conventional photographic techniques. Results may not be as "pretty" and sometimes need touch-up work. Most of the prototype boards were made in this manner. For copying to photosensitized boards, use figure 1-5*b*.

If you have a copier service make the transparency, tell them you need a reproduction with solid blacks. Lower densities where you can see through the dark areas will work if that's the best you can get. The toner adhering to the transparency is melted while being pressed onto the board with a clothes iron. Let the board cool, then peel away the transparency. The toner is left on the board to resist the etching solution.

Here are a few tips to make things easier. Set the temperature control on the iron to linen and let it reach full temperature. Clean and roughen the board with fine sandpaper or emery cloth. Preheat the board. Place a sheet of paper between the iron and the transparency. I always put several copies of the pattern on one transparency so that I can try again if I'm not satisfied. Acetone removes the toner if you need another chance. Film made spe-

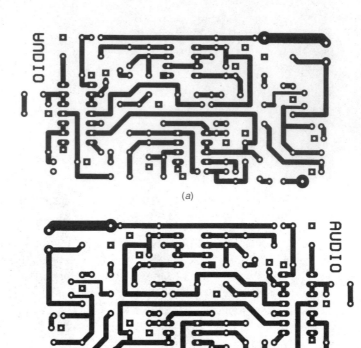

(a)

(b)

■ **1-5** *(a) The pattern for use with the transparency-film method. The ground plane faces the viewer; (b) The pattern for photosensitized boards (actual size). Square pads indicate connections to the ground plane.*

cially for this purpose is available.[1] Whether you use the special film or normal transparency stock, cut excess material away from the edge of the pattern. Material that extends ¼ in or more beyond the board edge tends to overheat, curl, and shrink.

The process described above does not always result in an unbroken resist pattern. One method I've used to compensate is to center-punch the board through the transparency while they are still stuck together after the ironing operation. With all drill holes marked, it's easy to touch up broken paths with a permanent ink marker.

After the board is etched and cleaned, you can start drilling holes for the component leads. The small square pads in figure 1-4 indi-

1 Tek-200 film from Ocean State Electronics, P.O. Box 1458, 6 Industrial Dr, Westerly, RI 02891, telephone (800) 866-6626, (401) 596-3080.

cate connections to the ground plane. Drilling them is optional because you can run the leads down through the holes or simply bend them parallel to the ground plane and solder. Countersink all of the holes on the ground-plane side except the ground connections. Some people prefer using a hand drill or a drill bit fastened in a control knob. I've had good results with a drill press running at 300 r/min or less. If you use a drill press, be careful. A ¼-in hole in the wrong place can ruin the board, not to mention your day!

I know from talking to other experimenters that some who enjoy most other aspects of building electronic gear really detest working with liquids used in etching boards, washing, etc. One way to construct most of the transceiver circuits is to use illustrations like figure 1-4 as a drilling pattern for single-sided copper-clad board stock. Use the copper-clad side as the ground plane. Countersink holes for ungrounded component leads, as in the previous method. On the unclad side of the board, bend component leads over and solder them to the appropriate points.

Check it out

Testing the audio board is a good idea, because if you know it's working correctly, it can help you test or troubleshoot the circuits discussed in chapter 2. You will need a well-filtered 12- to 14-V power supply. To activate the receive-mode amplifiers, connect a 5- or 10-kΩ resistor to pins 5 and 6 of IC1. Connect the free end of the resistor to supply. With a loudspeaker connected to C20, a faint hiss or white noise should be heard when volume control R18 is set to maximum. If you have an audio signal generator, a few millivolts of audio on C1 should produce a loud tone in the speaker. A less sophisticated but useful test is to touch the free end of C1 with a screwdriver or other metallic tool. A scratching or popping sound indicates amplification of the small change in potential as the tool touches C1. You will also probably hear a humming sound. This is caused by wiring in the house or building inducing small alternating currents in the tool. Grasping the metallic part of the tool or using an uninsulated tool will cause even louder sounds because of the pickup area of your body. (Microvolt currents in the 6-V range present at C1 are not dangerous. However, don't assume just because you can touch C1 with impunity that all low-voltage circuits are safe.)

To test the speech amplifier, connect pins 12 and 13 of IC1 through a resistor to supply, as mentioned above. Then disconnect the supply ends of the resistor controlling the receiving amplifier and connect it to ground. For this test you will need something to indicate the presence of an ac signal at C1. The best instrument for this is an oscilloscope, but an ac voltmeter can be used. Talking or whistling into the microphone should produce a signal at C1. With R6 set to minimum resistance, a loud whistle should result in a signal amplitude of more than 4 or 5 V peak-to-peak.

Double-sideband voice and CW modes

2

IN THE AUDIO CIRCUITS DISCUSSED IN THE FIRST CHAPTER, the goal was to amplify small electrical signals to form a larger replica of the original signal. Except for frequency-selective characteristics of the amplifier (roll-off), we don't want the waveform to be changed from its original shape. To do so will generate new frequencies. A misshapen waveform is usually known as *distortion*.

For instance, consider two pure tones, each a sine wave with a single frequency component. Both signals are applied simultaneously (with no interaction between the two) to the amplifier input circuit. The lower frequency could be 900 Hz and the higher frequency 1400 Hz. A perfect amplifier would have only these frequency components available at the output. In other words, one signal would not modulate or affect the other.

13

Real amplifiers may approach this ideal situation but fall short of achieving it, especially under certain conditions such as inappropriately high input levels. This can also happen to circuits other than amplifiers. Of the new frequency components generated, the two largest-amplitude frequency components are usually the sum and difference frequencies. For the audio frequency example above, the sum is 2300 Hz and the difference is 500 Hz.

Radio equipment often makes use of this process to shift the frequency of a signal. It is known by various names, such as mixing, modulation, or demodulation, depending on the application. A frequency-selective filter or other process can be used to select one of the new frequency components. When done correctly, the new signal is a faithful reproduction of the original but shifted in phase and frequency.

The main transceiver project is a double-conversion superheterodyne design. With two radio frequency conversions and one radio/audio frequency conversion, plenty of new frequencies or mixing products are generated. With so many signals, it's easy to end up with some in places that you don't want them. The end result can be spurious signals in reception and transmission. The diode mixers in this chapter do a good job of isolating each external connection or port from the other two to help reduce such occurrences.

Detector/balanced modulator

Many types of frequency-mixing circuits or mixers exist with varying properties. I decided to use a type known as a *dual-bridge diode mixer*. Other mixers can be used but you must first be sure you understand some of the basic requirements. Check some of the excellent sources listed in the Bibliography.

A dual-bridge diode mixer is shown in figure 2-1 and acts as the product detector for receiving and the balanced modulator for sideband transmissions. A lot of different circuits could be used here. I like this one because it works well with small-signal silicon diodes. Silicon diodes are plentiful and inexpensive. In single- and double-sideband transmissions, the steady RF carrier present in standard amplitude modulation is suppressed.

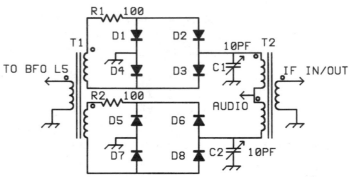

■ **2-1** *T1, T2—Amidon FT-37-43, five turns, trifilar wound. D1 through D8—1N4148 silicon or similar.*

The beat-frequency oscillator (BFO) amplitude is normally very low at the terminal labeled IF (intermediate frequency). Audio signals applied at the center tap of transformer T2 unbalance BFO currents in the diode bridges. Transformer T2 thus no longer has currents of equal and opposite phase in the primary winding. A double-sideband suppressed carrier signal is produced. Carrier suppression is −47 dB or better (at 4 MHz) in the prototype transceiver.

Another advantage of this circuit is it can handle lots of oscillator injection power. A large oscillator signal increases its ability to handle multiple high-level signals in receive mode with less inter-modulation distortion (IMD). IMD is like the unwanted mixing previously described. We want to generate mixing products from the BFO and input signal(s), not from two or more input signals mixing with each other.

The trimmer capacitors connected to each diode bridge can be used if you want to squeeze the last 2 or 3 dB of carrier suppression out of this circuit. The oscillator port has the highest return loss (lowest voltage standing wave ratio, or VSWR) when matched to a source impedance of about 100 to 150 Ω. Injection level is about 20 dBm. The IF port is closer to 100 Ω. If you substitute other mixers designed for 50-Ω systems, that's probably acceptable, although you may need to change the buffer output filter described in the BFO/Carrier Oscillator section.

Winding the mixer transformers

If you have previous experience building circuits similar to those in figure 2-1, feel free to skim ahead to the next topic. If not, transformers T1 and T2 may seem a bit bewildering. Their construction is really very simple when you understand the procedure—if procedure is the right word to use here. Once you have made some toroidal transformers, they won't seem so strange.

I'll get right to the point. Get some enamel-covered wire in the range of AWG 24 to 30. The choice of wire gauge is not critical. Cut it into three equal-length pieces of 7 or 8 in each. Color each end with nail polish or paint. Even better, use wire of the same gauge and different coloring, if you have some. The coloring is not essential but it helps you keep track of the windings, especially if this is your first transformer.

Lay the wires parallel to each other and twist them to form a tiny cable. The number of turns per inch is not really critical but try for about 5 to 7 turns per inch. Take the end of the tiny cable and pass it through the hole in the core. Bend it around the outside of the core and pass it through the hole again, as in figure 2-2. Do this five times while allowing the windings to travel around the core. The winding should spread out over most of the core. That's all there is to winding a trifilar toroidal transformer.

Now that the enigmatic little beast has windings, how do you connect it? The simplest way is to place it on the board, as shown in

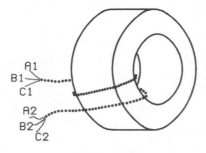

■ **2-2** *Each letter represents a winding. An ohmmeter will show continuity from A1 to A2, not from A1 to B1 or C1.*

figure 2-3. The traces on the underneath side are already configured to make the proper connections. Lines through the transformers, which connect holes on the board, represent a single winding. All you have to do is find two like-colored wires for each winding.

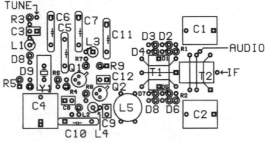

■ **2-3** *Some capacitors, such as C5, C6, may have different lead spacings, depending on the brand, type, etc. Just bend the leads to fit and don't allow the ungrounded leads to touch the ground plane.*

As an example, look at transformer T1. Suppose you have red-, blue-, and green-colored wires. On the left side of T1, insert the wires from one end of the cable. You could use the sequence red-blue-green, starting at the top. With the wires on the other end of the cable, insert them on the right side of T1 in the same sequence from top to bottom. Clean insulation from the wire ends, pull the wires snug, and solder them in place.

If you are making your own layout, refer to the phasing dots on the transformers in figure 2-1. Consider all of the dots to be at one end of the cable and undotted connections at the other end of the cable. One construction trick for keeping track of things is to tie knots in the wire ends represented by the dots.

BFO/carrier oscillator

The BFO provides the beat note for CW reception. It also is the signal needed by the mixer to demodulate SSB signals for reception. When transmitting SSB, its high-frequency signal is modulated to produce the initial double-sideband signal. A crystal filter is later used to remove one sideband.

Quartz crystal resonators used in the filter circuits (to be described later) are also used to control the BFO frequency.

The overall BFO circuit is shown in figure 2-4. In addition to the oscillator, a buffer and filter stage complete the circuit. Transistor Q1 is used in a variable-frequency crystal-controlled oscillator circuit. Q1 generates a signal at frequencies determined by Y1 and associated feedback components. Q2, the buffer amplifier, "steals" bias current from the emitter of Q1 and this common-emitter stage helps reduce effects on the oscillator frequency if load changes occur at T1.

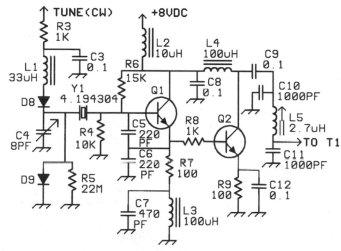

■ 2-4 *All 0.1-μF capacitors are monolithic ceramic types with 0.1-in lead spacing, and so are all other 0.1-μF capacitors in the transceiver unless specified otherwise. C30, C31—1.8 to 10 pF, Mouser stock number 242-1810 or similar. L5—2.4 to 4.1-μH variable inductor, JW Miller, Circuit Specialist catalog number 23A336RPC or similar. Adjust for output-voltage peak. Y1—4.194304-MHz quartz crystal, HC-49 holder. Mouser stock number 332-1042. Q1 and Q2—2N2222 or similar transistors. 2N3904 works well. D8 and D9—1N4148 silicon or similar.*

The reason for the output filter made of C10, L5, and C11 may not be obvious. The purpose is not really to reduce the harmonic content of the BFO. For best balance or carrier suppression, the diode mixer needs a symmetrical waveform at the BFO port. An output filter causes the BFO signal to more closely approximate the symmetry of a sine wave.

When used with the SSB filter discussed in chapter 3, the BFO output frequency needs to be about 4.1953 MHz. This circuit is designed to be able to oscillate at that frequency, even though the crystal is marked 4.194304 MHz. It can also be tuned to frequencies lower than 4.194304 MHz by applying a positive voltage to R3. Sending current through D8 effectively switches L1 into the circuit; the frequency of oscillation is changed. It can even be tuned over a small range for varying the pitch of received CW signals. The reason for this tuning range is to allow using a single, inexpensive, mass-produced quartz crystal instead of several custom-ground units.

When the dc path to D8 and D9 is opened, the frequency of Y1 shifts upward. With the components and board layout given, the upper frequency is around 4.1954 MHz and is used for generating lower sideband. The frequency of oscillation is about 4.193150 MHz when listening to a 600-Hz note for CW signals centered in the narrow IF filter. A mixing scheme in latter stages inverts the SSB signal when upper-sideband operation is needed.

An alternative oscillator

Y1 is operated both above and below the marked frequency on the holder. Because of the relatively low frequency, temperature-induced frequency drift has not been a problem. However, it's not as stable as a circuit using correctly ground crystals with the manufacturer's specified capacitance. Some versions of this circuit may refuse to start consistently. You may have to adjust the value of R6 to compensate for this tendency if it occurs. If you plan to operate the rig in extreme conditions, switching between two crystals is a possibility. See figure 2-5.

Using custom-ground crystals with the correct parallel capacitance is a reliable method of putting the oscillator on frequency. The circuit layout will be slightly different from those shown in figures 2-3 and 2-6, but not enough to mandate a completely new layout. Judicious use of a small relay, glue, and a razor knife is sufficient to modify the board.

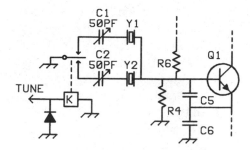

■ **2-5** *Almost any silicon diode is suitable here. The tune terminal is supplied by a circuit that can easily be modified to provide coil voltages up to 12 V, so use a small single-pole, double-throw relay of your choice. Quartz crystals are custom ground for the frequencies indicated.*

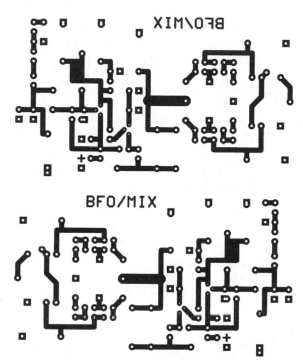

■ **2-6** *Remember that the pads and paths with square corners mark grounded connections and must be connected to the ground plane on top (actual size).*

I suppose I should explain what all this verbiage about custom-ground crystals and mass-produced crystals is referring to. *Custom* means just what it implies. You send in an order to the manufacturer stating the specifications for the crystal (within limits) and they fabricate one for you. Some manufacturers are happy to fill small orders, and these manufacturers are listed in the Address section at the end of the book.

The frequency I chose for this project probably seems rather strange. You may be wondering why anyone would make lots of quartz crystals for such an odd frequency. The reason is that 4,194,304 is equal to 2 raised to the 22d power. If you take 4,194,304 Hz and divide it by 2, divide the result by 2, and so on 22 times, you end up with 1 Hz. A lot of applications exist for timing events in one-second increments, so that is why we have the crystals available with a seemingly strange resonant frequency.

Construction techniques for this board as well as the other transceiver boards are similar to the methods used in the audio board. Mount transformers T1 and T2 on edge with the toroids standing up and supported by six wire leads. Mount L6 by the leads or by drilling a hole slightly larger than the coil form. Press the form in, cement it in place, and wire the terminals to the board with short lengths of solid hookup wire.

Figure 2-6 is the resist pattern. After you have etched, dried, drilled, and countersunk the board, the components can be inserted. Most of them are easy to place. The cluster of diodes around T1 must be inserted with the proper polarity, which can be a bit confusing. Here's a simple tip to make it easier.

Remember that the band around the diode body represents the cathode. That's also the bar on the schematic symbol pointed to by an arrow. The group of diodes above T1 (D1 through D4) should have their bands away from R1. If these diodes have symbols on them, the arrows will all point away from R1. The group of diodes below T2 point in the opposite direction, toward R2.

Keyed carrier generator

Continuous-wave telegraphy transmissions can be generated by several methods. Applying an audio sine wave to the audio port at T2 will generate the equivalent of two continuous-wave carriers. A selective filter can then be used to eliminate one of them. Many commercial transceivers have used this method. Making a system like this, which offers a tunable BFO in receive mode along with a tracking side-tone, can be a challenge. It's not hard to build a good sine-wave oscillator, but making one that covers a wide frequency range is more difficult.

Another method is to apply some direct current to the balanced modulator to unbalance it. The BFO signal then comes through to be processed by the rest of the transceiver circuitry before going to the antenna. In normal operation, two or more stations communicating with each other will transmit on the same frequency. If the same BFO frequency is used for transmit and receive operation, no audio tone or note will be generated in the detector. The BFO or some other oscillator in the transceiver will have to shift frequency slightly to receive other on-frequency signals.

The method used in this chapter for generating a CW signal is simply another oscillator operating in the IF passband. The schematic diagram is shown in figure 2-7. Think of it as a one-transistor, low-power transmitter. A signal from this keyed oscillator mixes with the BFO to produce an audio tone just like incoming signals do when receiving. The oscillator should be tuned to the center of the narrow IF filter passband. Operators usually expect to hear received signals with a pitch similar to the monitor side-tone generated when keying. Side-tone pitch automatically tracks the pitch of incoming signals even if you retune the BFO while listening to a CW signal.

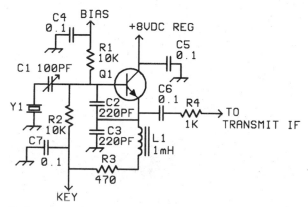

■ **2-7** *R2 is returned to the key line. This prevents exceeding the base-emitter reverse breakdown voltage of Q1 when bias is zero. C1—Digi-Key part number SG3016-ND or similar. C2 and C3—Zero-temperature-coefficient ceramic or other temperature-stable type. Q1—2N2222 or similar.*

The terminal labeled "key" is connected directly to the key jack. Bias for this oscillator comes from a control circuit and is used to enable or disable the oscillator, depending on the mode selected. Resistor R4 attenuates the output to a level suitable for injecting into the transmit IF circuits.

Component locations for the oscillator appear in figure 2-8. Connecting R4 is probably best done with miniature coaxial cable.

Using unshielded wire may cause problems with instability because energy may be radiated or picked up from other parts of the transceiver. It's also a good idea to install the board so that the cable connecting the IF circuits is short. Etching patterns are shown in figure 2-9.

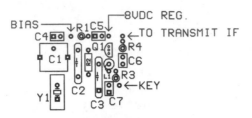

■ **2-8** *This layout shows a plastic-case transistor. Metal-case types are all right too, but be careful to insert leads in the right place.*

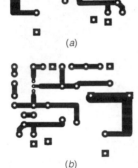

(a)

(b)

■ **2-9** *Transistor lead spacing is pretty close. Check Q1 connections on the finished board for shorted connections. Use the transparency method for (a) or the photographic method for (b) (actual size).*

On-the-air already?

The boards previously discussed can be used in several ways. If you build the audio and BFO/detector boards, guess what? You have a working receiver. The BFO frequency is already close to the 75-m phone band. If you wish, change the crystal to an amateur frequency and make the changes shown in figure 2-10. The frequency of a crystal resonator cannot be moved very much. Don't expect more than 4 or 5 kHz of tuning range at this frequency. If you want to monitor a regularly scheduled net or roundtable, this may be sufficient.

This kind of receiver is useful for listening to SSB and CW signals. Signals on either side of the BFO frequency are detected equally

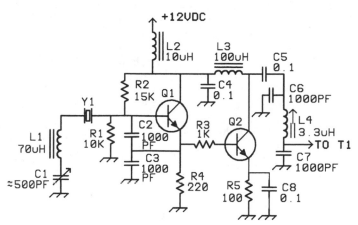

■ 2-10 *It takes a relatively large change in reactance to pull a low-frequency crystal such as Y1 several kilohertz. The value for L1 is a suggested value and will depend on the individual crystal.*

well. It does not have the kind of selectivity that other more sophisticated receivers have. Although lacking in a number of features, receivers like this are certainly useful. I have personally made numerous on-air contacts using a direct-conversion receiver or transceiver.

Top end for the tuning range will be approximately the frequency marked on the crystal holder. Tuning range depends on the type of crystal. Usually those crystals in the larger holders can be pulled most readily from the intended frequency. Crystals in the FT-243 or HC 6 holders usually work better in this circuit, as opposed to smaller types such as an HC-18.

Coverage of the entire 80/75-m amateur band would certainly require a large number of crystals. An LC resonant circuit for tuning is a more practical solution (see figure 2-11). The value of C1, C2, C3, and L1 determine the frequency of oscillation. Stray capacitance and internal capacitance at Q1 also have a small effect. Starting at 4 MHz, this circuit will tune downward more than 300 kHz.

Why even bother with a crystal resonator when a simple LC resonator can be adjusted to cover whatever frequency you desire? The main reason is that quartz crystal resonators are inherently more stable. Their resonant frequency is less affected by external influences. Great care is needed in the design and construction of LC tuned oscillators to achieve frequency stability approaching that of crystal oscillators.

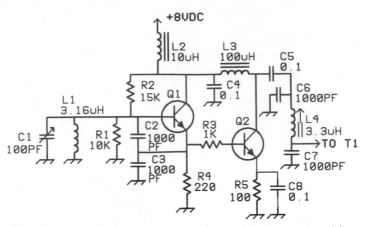

■ 2-11 *This circuit has a much wider tuning range and is a good choice for the direct-conversion receiver.*

Fortunately, factors affecting frequency stability are not quite as hard to control for this frequency as would be the case at higher frequencies. Using a regulated power supply helps. The detector presents a reasonably steady load for the BFO and a buffer stage also aids stability. Temperature changes cause some components to change value slightly, causing a slow drift in frequency. To combat thermally induced drift, use zero-temperature-coefficient capacitors in the resonant circuit. Use an air-core inductor for L1.

Receiving applications are perhaps less stringent than trying to maintain a steady frequency when transmitting. Usually fewer conditions exist to upset a variable-frequency oscillator. Conditions such as more heating, sudden load changes, and strong RF fields must be dealt with in transmitters. Operators can often manually retune a receiver as a last resort to compensate for instability.

If you feel as though the utility of an expanded tuning range is worth the effort, this is a good frequency range to learn from. It is low enough to be practical instead of exasperatingly difficult. It is also high enough to demonstrate the effects mentioned earlier. More information about oscillator stability is presented in later chapters.

Figure 2-12 shows how to use the circuits covered up to this point as a direct-conversion receiver. Signals from the antenna are mixed with the BFO. Difference (and sum) frequencies are generated. Signals very close in frequency to the BFO result in mixing products in the audible range. These are amplified in the audio board. There is no amplification between the antenna and detector as in larger receivers. If you try this circuit, sensitivity can be increased by changing R7 in figure 1-2 to 100 kΩ.

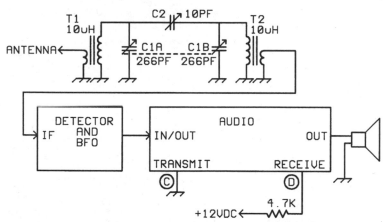

■ **2-12** *T1 and T2 transformers made from Amidon T50-6 cores 50 turns of AWG number 28. Each link winding is five turns. C1 is a dual gang variable capacitor Mouser part number 24TR218.*

A manually adjustable bandpass circuit is used to limit the number of signals reaching the detector. This helps to prevent AM broadcast signals from overloading the detector and interfering with desired signals. Wind transformers T1 and T2 in the same manner as the mixer transformers, but do not twist the wires together before starting the winding. Make two separate windings on the core.

Tune C1 for maximum signal strength on a loud, steady signal. Adjust C2 toward minimum capacitance. Spread or compress turns of the larger winding on T2 for maximum signal strength. Retune C1 and adjust T2. Now adjust C2 toward maximum capacitance and a louder signal. Tuning C1 should show a definite peak in loudness when receiving signals in different parts of the band. If you notice two pronounced peaks for each signal, reduce C2 a bit.

The simple receiver is fun to listen to and sounds really good under favorable conditions. How about transmitting? After all, the boards are designed to transmit as well as receive signals. Figure 2-13 shows how to make a double-sideband transceiver. The 4-MHz low-pass filter and amplifier are described in chapter 5. Switch S1 and relay K allow the operator to change between transmit and receive modes.

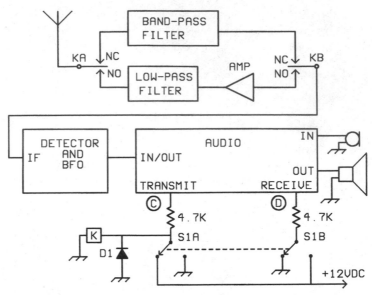

■ **2-13** *A simple double-sideband transceiver. It's easiest to use one of the crystal oscillator circuits for the BFO. Experienced builders familiar with shielding, power regulation, and mechanical stability requirements may want to try using figure 2-11 for the BFO.*

When the operator makes a voice transmission with this system, sidebands above and below the BFO frequency may be picked up by other receivers tuned to the same frequency. Most HF receivers in use today can eliminate one of the sidebands with internal filtering. If both sidebands are allowed to reach the product detector, they combine in a destructive way that produces unintelligible audio unless the BFO of both transmitter and receiver are on exactly the same frequency and in phase.

To put it another way, communication between two direct-conversion DSB transceivers is not practical. You can talk to stations using conventional SSB equipment, however. Another issue is the extra spectrum occupied by such a signal. Twice the band space needed by an SSB signal is occupied. Using this mode with high power and crowded band conditions is not a good idea. If you try this circuit, keep these considerations in mind.

The simple double-sideband receiver or transceiver is fun to use; however, you may want to build two sets of boards. The rest of the circuitry in subsequent chapters is easier to test and align with these stages connected and working.

Touchy tuning

Some builders may want to make this chapter's projects permanent rather than treating them as experiments or building blocks of a larger transceiver. Tuning SSB and CW signals requires a slow tuning rate because the pitch of recovered audio changes as the BFO frequency changes. With the variable crystal oscillator (VXO) circuit, tuning rate won't be a problem because the entire sweep of capacitor C1 will cover only a few kilohertz.

Tuning a few hundred kilohertz with the circuit shown in figure 2-11 is a different matter. Simply mounting a knob on the capacitor will work, but it quickly becomes annoying to attempt rotating the tuning knob in such fine increments. A simple means of slowing the tuning rate is needed. Perhaps the simplest method is to install an extra variable capacitor. It should be much smaller than C1 with a maximum capacitance of 5 to 10 pF. The new capacitor is used as a fine-tuning control.

The two-capacitor arrangement helps, but it is a cumbersome way to scan the band. From an operator's viewpoint, mechanical reduction is probably a better solution. Gears and/or pulleys can be used to slow the rotation of C1 as the main tuning knob rotates. Geared knobs or drives are generally not so available as in the past. Of those that are, inexpensive ones perform poorly, while quality units are too expensive to justify for small projects. An exception is the high-quality drive salvaged from surplus equipment, if you are fortunate enough to find it.

Pulley-and-cord mechanisms have been used in many commercially produced radios. If carefully designed, they are an excellent method of drive reduction for home-built radios. I have constructed some in my own equipment which have no detectable backlash. Another advantage is that the panel and capacitor are physically isolated from each other. Disturbances at the panel or cabinet don't affect the tuning as much as some geared or vernier drives do.

The easiest way to get started is to use a salvaged pulley from a defunct radio. Suitable mode range from tube types to present-day receivers. If you don't find e size you want, try making a pulley. The pulley in figure 2-14 is made from a metal lid liberated from a mayonnaise jar. Threaded grooves on the edge of the lid act as guides to keep the dial cord in place. A flat-sided aluminum knob is bolted to the center and forms the hub.

27

■ 2-14 *A home-built cord-and-pulley reduction mechanism results in smooth, easy tuning. You will have to determine for yourself whether to purchase low-fat or regular sandwich spread.*

It's important to center the hub as accurately as possible. One good way to do this is to use a bisecting square. Draw or scribe a line along the 45° straight edge. Rotate the lid about 90° and mark another line. The lines will cross at the center of the lid. You may want to mark one or two additional lines at different angles to double-check the position. The same method can be used to find the center on the knob.

Dial cord usually has a nonstretching inner core of glass fiber and an outer jacket of a durable synthetic fabric. Unlike pulleys used for power transmission, both ends of the cord or belt are usually anchored to the pulley. Both ends of the cord are wrapped once or twice around the pulley before entering a hole or slot on the outside guide groove. Each end enters the hole from opposite directions. To eliminate slack, small coil springs are used inside the pulley. One end of the dial cord is tied to the spring, which is anchored inside the pulley. The other end is either anchored or tied to another anchored spring.

The hole or slot allowing passage of the dial cord must be smooth and free of sharp edges so as not to damage the cord. A knob, shaft, and bearing assembly are needed to complete the mechanism. The cord is usually wrapped twice or more around the shaft to prevent slippage. Most commercial implementations use a groove or guides on the shaft to keep the cord from wandering or creeping along the shaft as it rotates. I've found that this is not always necessary. A straight ¼-in shaft often works well, even if the dial cord wanders along the shaft a bit.

Low intermediate-frequency circuits

THE SIMPLE RECEIVER AND TRANSMITTER OUTLINED IN chapter 2 detected or generated RF signals directly at the operating frequency. This chapter shows how to use those circuits along with new ones to build a portion of a more sophisticated radio transceiver.

A fixed intermediate frequency allows us to do things to the signal that would be impractical over the frequency span used in the preceding chapter. It is much easier to build highly selective filters for a single frequency than to accomplish the same thing with tunable filters, for instance. Amplification and gain control are also simpler.

Receive-mode IF signals

The circuit in figure 3-1 performs several functions. It amplifies weak signals. It also provides some filtering and impedance matching. Capacitor C1 connects directly to a crystal filter for SSB. As we will see later, CW signals are also routed to C1 through a matching circuit.

A field-effect transistor (FET) at Q1 is used to drive IC1. The crystal filter modules have their own built-in terminating resistors and switching arrangement. Each filter is connected or taken out of operation by the activation of external control lines. The impedance for the source follower is high, so it does not affect the selected filter module. IC1 and IC2 provide lots of gain and drive a filter which includes Y1 and Y2. Y1 and Y2 help reduce the amount of wideband noise generated by the amplifiers and sent to the detector. Another FET, Q2, drives the detector.

Locations labeled Gain1 and Gain2 are used to control amplifier gain. Minimum gain occurs when a positive voltage of 7 to 12 V is

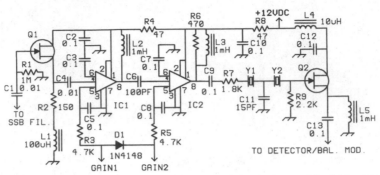

■ 3-1 *During reception, incoming signals are amplified by this circuit. Q1 and Q2 are 2N5485 or similar. IC1 and IC2 are MC1350P. Crystals ground for 4.194304 MHz are Mouser part number 332-1042.*

applied to pin 5. Reducing the voltage increases the gain. Maximum gain occurs below approximately 5 V. A potentiometer or automatic gain control circuit can be connected to Gain1 and Gain2, effectively eliminating D1. If you use a single panel-mounted potentiometer, adjustment is rather touchy. A better solution is to use a trim pot (board-mounted potentiometer) for Gain1. The trim pot will need to be set to somewhere near maximum gain for best weak-signal reception but not so high as to overload IC2 on strong signals.

Presetting Gain1 allows manual or automatic gain control to be achieved more smoothly. Connect the trim-pot wiper to Gain1, ground to one leg, and an 8-V regulated supply to the other leg. A control circuit, to be described later, also sends a positive voltage to Gain1 and Gain2. This control signal turns the amplifiers off to keep transmitted signals from circling around the circuit shown in figure 3-1 in a feedback loop.

No automatic gain control (AGC) circuits are present in the transceiver at the current stage of development. This is not because of any inherent shortcoming in the design. I simply devoted more time to critical (challenging?) aspects in this project. I personally don't find the lack of AGC bothersome; however, I'm aware that some operators certainly will. Sources listed in the Bibliography provide information and examples of automatic gain control.

The easiest place to sample incoming signals for an AGC circuit is probably the audio power amplifier output. The signal amplitude is so large that less circuitry will be needed to process it into a control signal. Audio-derived AGC is relatively simple. Sampling a signal from the IF chain can provide better performance, although it may be slightly more complicated.

Transmit-mode IF signals

A double-sideband suppressed carrier signal is generated by the balanced modulator shown in figure 2-1. Another 4066 electronic switch, shown in figure 3-2, gates signals on and off. A DSB signal entering at pin 8 is allowed through switches C and B all the way to pin 3 when pins 5 and 6 are held high by the control circuit. A circled C indicates one of the control-circuit connections. At the same time, switches A and D are open because of a low generated by the circled-D control-circuit connection.

Transistor Q4 is part of a common-emitter amplifier stage with a fixed ac impedance path from emitter to ground. The dc path from emitter to ground includes a variable resistance, R19. Adjusting resistance in this path allows us to change the point where Q4 collector current saturates. Two limits are established for the output waveform of Q4. One is this saturation point beyond which collector current cannot increase. The other is collector current cutoff. A signal which drives collector current between these two points is amplified in a normal manner.

Voice signals have high peak amplitudes compared to the average amplitude in a waveform. By limiting the peaks, you can increase the average power in a signal. This may be helpful under some conditions. If your signal is near the noise level for a distant receiving station, only the peaks may be coming through on a nonlimited signal. Peak limiting allows you to turn up the drive control a bit

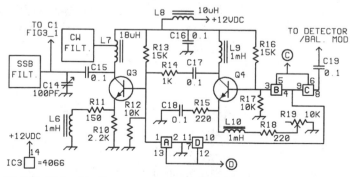

■ **3-2** *This circuit accommodates filter impedances, switches signals, and clips voice peaks. Q3 and Q4 are 2N2222 or similar. IC3 is a CD4066 or equivalent. C14 is a Digi-Key part number SG3011-ND.*

without overdriving subsequent transmit amplifiers. Failure to limit the peak amplitude will cause distortion and a widening of the signal bandwidth known as *splatter.*

The limiting action also causes distortion of DSB and a widened signal. Spurious products generated by limiting are removed by the crystal filter, which also helps remove some of the audible distortion. Up to a point, the limiting can improve intelligibility under some conditions. Too much limiting in this stage may cause enough distortion to make speech less understandable on the receiving end. Mounting R19 on the front panel is the easiest way to experiment with different settings. If you do this, it's a good idea to prevent unwanted feedback with a 0.1-μF capacitor from ground to the junction of R18 and L10.

This kind of limiter or clipper circuit seems to sound less noisy than some other schemes which clip the audio signal instead. A better implementation might use an additional crystal filter before Q4. In any case, the results depend to some extent on band conditions, receiver characteristics, and operator preference at the distant station. When signals from the other station in a QSO are strong, some operators seem to prefer a signal with little or no processing.

Next comes Q3, another common-emitter stage with a different biasing scheme. The main purpose of this stage is to facilitate impedance matching and provide a low but predictable amount of gain. AC impedance from emitter to ground is determined by R10. A much lower dc path through R11 and L6 allows enough emitter current flow for good linearity. C14 and L7 in the collector circuit are tuned to resonate at the IF frequency. The DSB signal is sent to the SSB crystal-filter module.

The CW-filter module is not used for transmitting. The reason for connecting to a tap on L7 is that C14 and L7 are also used in receive mode. Q3, however, is turned off. When receiving, switch A is turned on, bringing the base of Q3 close to ground potential. With Q3 effectively out of the circuit, the low-impedance output of the CW filter is matched to the high-impedance gate of Q1 shown in figure 3-1. An oscillator circuit shown in figure 2-7 is used for generating CW signals when transmitting.

Although the circuit shown in figure 3-2 is used to process signals when transmitting, it has a very important side effect during reception. It effectively provides a path from the output of a high-gain receive IF amplifier back to the input. When transmitting, receiver amplifiers as in figure 3-1 are disabled, so there is no problem. During reception, however, attenuation with the circuit shown in figure 3-2 must be very high, otherwise unwanted feedback and oscillations will take place.

More matching and amplification

For the moment, let's assume the crystal filters are already in place. Figure 3-3 shows the circuit connected to the other end of the filters. When receiving, Q5 is active, while Q6 is turned off. The circled-A connection is grounded by the control circuit and the circled B connection is forced high. The situation is reversed when transmitting. L12 and C23 are tuned to resonate at the filter passband frequency. This circuit is also used by Q6 as an input tuned circuit.

Circuits shown in figures 3-1 through 3-3 all work closely together, so I put them all on one board layout, as shown in figure 3-4. The two large rectangles with no component outlines are spaces reserved for the crystal-filter modules. You will find it easier to place and solder components with ground-plane connections first. Other components with no grounded leads will not be in the way

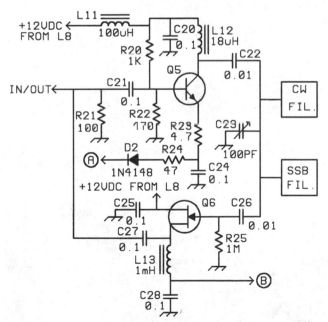

■ **3-3** *This circuit interfaces with the other end of the crystal filters. Q5 is a 2N3904 or similar. Q6 is a 2N5485 or similar. C23 is a Digi-Key part number SG3011-ND.*

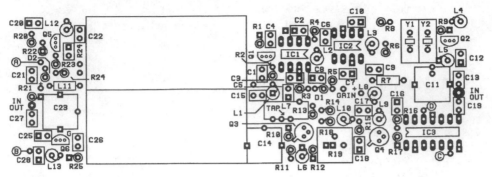

■ 3-4 *This board processes transmitted and received signals at the lower IF frequency.*

as you try to place iron and solder as needed. You can subsequently stuff other components, bend leads, and flip the board over and solder them on the etched side of the board. Etching patterns for the circuits shown in figures 3-1 through 3-3 are shown in figure 3-5.

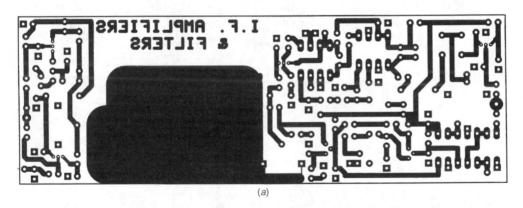

(a)

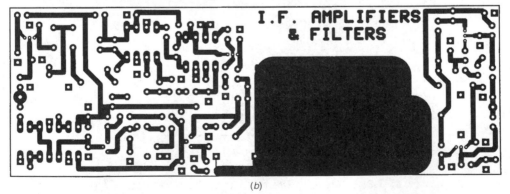

(b)

■ 3-5 *Etching patterns (actual size) for the circuits shown in figures 3-1 through 3-3: (a) for reproduction by the transfer method and (b) for use with photoetching methods. Remember to check for shorts between closely spaced paths.*

Crystal filters

Two ladder-type six-crystal filters are used to select desired signals in the IF passband. Figure 3-6 shows the SSB filter along with components used to switch it in and out of the circuit. It also contains terminating resistors, and I'll refer to the whole assembly as a *filter module*. When a filter module is inactive, control is at ground potential. This biases Q1 and Q2 to pinch off any channel current and stops signals from flowing through the FETs. Control is brought to a positive 12 V to allow signals through.

Bandwidth for this filter is about 2700 Hz at 6 dB down from the peak response.

Capacitors C8, C9, C11, and C12 are shown as variable capacitors. The value listed for each capacitor is not the maximum, as is customary for labeling variable capacitors. It is the actual value to which each capacitor is adjusted. I used trimmers in the prototype; however, the values listed are available as standard fixed units. Using these values should result in a reasonably good passband response.

To do an even better job, some alignment equipment will help. If you are lucky enough to have access to a spectrum analyzer and tracking generator, have at it. Chapter 18 describes an inexpensive project made for use with most oscilloscopes. It will allow you to see the actual passband shape.

Another almost identical filter module is shown in figure 3-7. All of the connections are the same and a few parts have changed value. The 6-dB bandwidth is about 600 Hz. Terminations for this filter are much lower—240 Ω. It's not as touchy as the other filter. Simply use fixed capacitors with a 5 percent tolerance rating. It has more insertion loss and the passband response is not as flat.

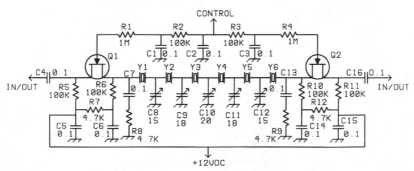

■ **3-6** *This is the wide SSB filter. Q1 and Q2 are 2N5485 or similar. All crystals are 4.194304 MHz, Mouser part number 332-1042.*

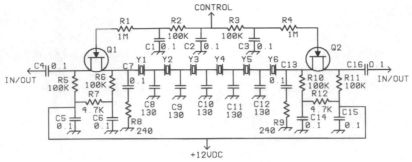

■ **3-7** *The narrow CW filter uses the same layout and wiring as that in figure 3-6. Only the part values have been changed.*

To house the filter modules, I used small, stamped-metal boxes from Mouser Electronics. Component density inside the boxes is very high—at least by most hobbyist standards. If you wish to enlarge the boards shown in figure 3-8, larger enclosures can be accommodated by modifying a paper copy of the patterns shown in figure 3-5 and adding area to the rectangles reserved for filter modules.

If you do use the small boxes, here is how the boards are mounted. Boards depicted in figure 3-9 have insufficient space for mounting screws. Enough space exists along the longest sides to solder some thin brass or copper sheet to the ground plane. After these tabs are soldered in place, bend them upward to form a 90° angle with the board. This will allow you to insert the board into the box, etched side first. It should go in just far enough to bring the tops of the crystal cans below the opening edge of the box. Drill two small holes on the box sides near the opening edge where the mounting tabs lie. You can mark the tabs or very carefully drill through them at this point. Use care to prevent the bit from cutting into a crystal or other component if you drill the box and tabs at the same time.

Another way to fasten tabs to the box walls is by soldering along the edges. Don't use too much heat and solder in case you want to

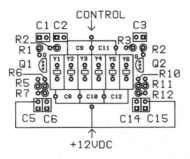

■ **3-8** *C8 through C11 are soldered to the ground plane where convenient and depend on what capacitor style is used. A trimmer such as a Mouser part number 24AA023 will fit. The following components are not shown because they are soldered in place and supported by their leads: C4, C7, C13, C16, R8, and R9.*

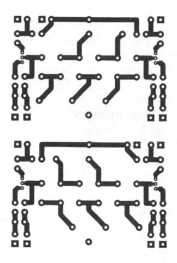

■ **3-9** *These are really small boards. Crystal cans are soldered to the ground plane at the end of each can (actual size).*

remove the boards in the future. Figure 3-10 shows how the modules look with the boards nestled in place.

Construction and layout tips

Once the board shown in figure 3-4 is completed and inspected, a nonconductive table or desktop is fine for initial testing. Connect

■ **3-10** *An example of a filter module installed in a metal box.*

the in/out terminal where C13 and C19 join with miniature 50-Ω coaxial cable. Use a few inches of cable to connect the detector/balanced modulator at the other end. Temporarily connect a 50-Ω resistor (don't use a wirewound resistor) to the other in/out terminal. Other connections can be made with standard hookup wire.

For testing purposes, the audio, BFO, and lower IF boards should all be connected appropriately and spread out over the nonconductive work area. Testing procedures will be covered shortly. For now, this is your temporary layout. When you assemble these boards in close proximity to each other, certain precautions must be taken to prevent degraded performance. One easily observable effect is the BFO radiating energy which is picked up and amplified by IF amplifiers. This causes reduced carrier suppression when transmitting SSB signals. It also degrades reception by modulating incoming signals with noise.

In a later chapter I will cover more ideas for packaging the circuits. It's not too early to start thinking about how you want to package the various transceiver boards. One possibility is to mount everything breadboard-style on a large piece of wood or aluminum or unetched printed circuit board. Everything is accessible as you keep adding circuits to the transceiver. A wood support should be covered with copper foil or flashing to simulate the continuous ground plane of a metal cabinet. After the troubleshooting and testing is over, it's all ready to be mounted in a permanent enclosure.

Testing the SSB generator

At this point in the project, control circuits are not ready, so a few temporary connections are necessary for testing. For SSB transmissions, an oscilloscope can be used. Connect the scope probe to the temporary 50-Ω resistor at the in/out terminal. Connect all points requiring positive 12-V dc to a well-filtered, regulated supply. Connect Gain1 and Gain2 to 12-V dc. Connect circled C to a resistor of about 5 or 10 kΩ to 12-V dc. Connect circled D through a similar resistor to ground. Circled B goes directly to ground and circled A directly to positive 12-V dc. For the filter modules, supply both with positive 12-V dc and connect control on the SSB filter module to positive 12-V dc. Connect control on the CW-filter module to ground.

Whoa! That's a lot of connections. When all of the control circuitry is in place it will do much of this automatically. Adjust R19 for minimum resistance. Make sure the audio board is set up in the transmit mode. Apply a low-distortion source of audio tones to the microphone input of the audio board. A two-tone generator, described at the end of this chapter, is one way to do this. If you have nothing else, plug the microphone in and whistle into it. A clean, high-pitched whistle closely approximates the shape of a sine wave. The oscilloscope trace should fatten into a wide bar.

After the crystal filter removes one of the two frequency components of a DSB signal, the scope trace of an SSB signal generated by a single audio tone looks like a straight bar. If anything other than a single frequency is present in the output, the edges of the bar will be rippled, as in figure 3-11. This can be caused by several things, such as carrier leakage, insufficient sideband rejection, hum, distortion in the audio generator, or distortion of the DSB signal before it reaches the crystal filter.

You can use the amplitude of ripple shown in figure 3-11 to determine how well the upper sideband is suppressed. If none of the other factors mentioned in the previous paragraph are excessive, here is how to interpret the scope trace. On the scope graticule, read the peak-to-peak variation of ripple along one edge. Compare this to an averaged peak-to-peak reading for the entire waveform. A 10-to-1 voltage ratio represents 20 dB of suppression; 100-to-1 represents 40 dB.

$$\text{Suppression} = 2(\log \text{ voltage ratios})(10) \qquad (3\text{-}1)$$

Obviously, when the ripple amplitude is 100 times smaller than the full waveform it becomes difficult to make accurate measure-

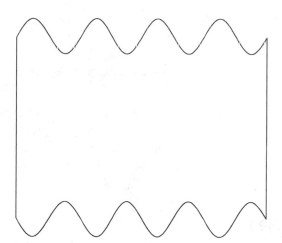

■ **3-11** *Amplitude modulation or ripple caused by other frequency components sneaking in on your hoped-for single-frequency signal.*

ments. You can increase the resolution of your measurements to some extent by switching to different vertical deflection sensitivity settings. In any case, many experimenters own or have access to an oscilloscope. This is a quick and easy test in lieu of using more sophisticated and expensive equipment.

A general-coverage communications receiver is useful for listening to the output. It's best to have someone else talk or patch in canned audio from something like a tape recorder. Listen to the communications receiver with headphones to check the signal quality. You can also determine to some extent how good the opposite sideband suppression is. The results will probably not be very accurate unless your assembled boards are shielded from the receiver.

Testing SSB reception

Checking out the receiver circuits requires changing a few connections to control points on the assembled boards. Refer to chapter 1 to set up the audio board. Make the following connections to the IF board: circled C to ground, circled D to supply, circled A to ground, and circled B to supply. Connect Gain1 and Gain2 to a variable voltage source or potentiometer.

You will need some kind of signal source. If you have a conventional signal generator, connect it to the in/out terminal at C21. Sensitivity at the in/out terminal can be very high, depending on the voltage at Gain1 and Gain2. This high sensitivity can be used to advantage if you don't have a conventional signal generator. One or two feet of hookup wire connected to the in/out terminal will act as a small antenna. The radiated signal of a dipmeter is usable in this manner. The local oscillator from some portable shortwave receivers is also useful.

If you can hear test signals and everything seems to be working, try the narrow CW filter. Swap ground and supply connections for control wires connected to the filter modules. Tuning of test signals should appear much sharper. You should hear the test signal over a much smaller frequency span.

Alternatives

At this point, I'm sure some of you more accomplished experimenters are seeing instances when you would do things differently. Good! There are certainly many ways to make an IF strip. I hope you find some of the preceding ideas useful. With separate boards, there's no reason why you shouldn't mix and match, using your own ideas along with those you find suitable here. Input and output connections are low impedance and suitable for miniature 50-Ω coaxial cable.

Other IF frequencies could be used here. Instead of 4.2 MHz, you could use a higher IF and commercial crystal filters. The high IF is 72 MHz, so you might want to stay away from a submultiple such as 9 MHz. Separating the transmit and receive chains completely is another possibility. You could make almost all stages dedicated to receiving or transmitting with only common oscillator circuits.

A two-tone audio signal generator

Figure 3-12 is the complete schematic for producing two independent, spectrally pure sine waves. The signals are combined in a manner that allows adjusting the amplitude of one frequency component relative to the other. In addition, the composite signal amplitude can be changed. Either oscillator can be turned off if a single frequency is needed.

The oscillators use an LC tuned circuit. Most audio generators for transmitter testing use carefully designed RC feedback networks. Every time I looked at the surplus 88-mH toroids in my junk box, they appeared more obsolete. Although the toroidal inductors can be used in audio filters, other technologies seem more appealing.

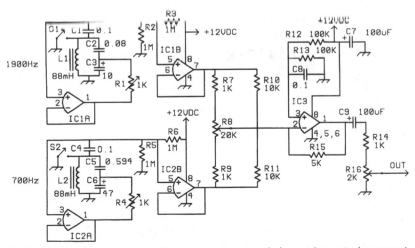

■ **3-12** *The frequencies of signals generated do not have to be exact; however, they should not be harmonically related.*

First it was active RC filters, then switched-capacitor filters, now digital signal processing. Rejoice toroids! Now you have a purpose. Oh . . . I mean, this seems like a good place to apply the toroids.

This design may weigh a bit more than other generators and maybe that makes it a bit dated (remember that next time you go backpacking with a two-tone generator). It is, however, very forgiving of component values. The inductors are used as a filter in the feedback network of an oscillator. The filter components are not especially critical. As an example, the value of C2 can vary a bit with a small change in frequency. A similar percentage change in C3 will have an even smaller effect. In either case, you can still adjust R1 to produce a low-distortion sine wave.

With both oscillators running, tests on a digital spectrum analyzer show no spurious signals greater than −45 dB. Spurs above 3000 Hz, which could cause out-of-band emissions and single-tone harmonics, are even lower in amplitude. Other inductors may be used. Even with lower-Q inductors, the waveform distortion is low enough for meaningful transmitter test. R1 and R4 are adjusted to produce 2 to 3 V peak-to-peak at the output jack with RB centered and R16 at maximum output.

The prototype was built on perforated board stock. Placement of components and routing is not critical. Figure 3-13 shows the parts layout diagram for a printed circuit board. Mount the toroids with a bolt through the center. Use large insulating washers to protect and insulate the windings. Locations for C2 and C5 show three

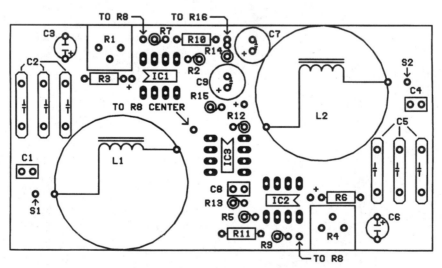

■ **3-13** *A suggested layout for the two-tone audio generator.*

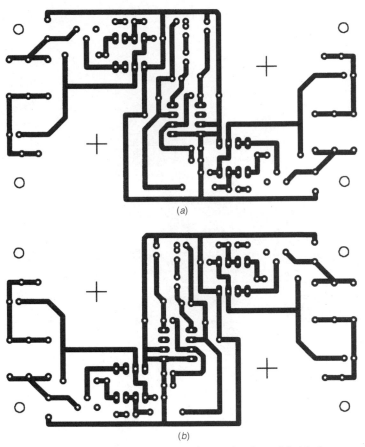

(a)

(b)

■ **3-14** *(a) is for reproduction by the transfer method, and (b) is for use with pho-toetching methods. Alternatively, you could use (b) as a wiring plan for a perfo-rated board (actual size).*

capacitors. The reason for this is to allow placing combinations of capacitors in parallel to achieve a desired value.

Mounting centers for L1 and L2 are marked in Figure 3-14 by crosshair symbols in the etching pattern. Single-sided stock is suitable for this board because a large low-inductance ground plane is not necessary. This board should not be mounted in the same box as an ac power supply unless you keep it well away from the power transformer.

High intermediate-frequency circuits

4

QUITE A NUMBER OF CIRCUITS HAVE BEEN ILLUSTRATED at this point. Next comes an overall look at where we are going with all of this. Hopefully, familiarity with the earlier chapters will be a point of reference from which to view the tasks which lie ahead.

Most circuit boards for the rest of the transceiver are constructed in the same manner. If you are making successful circuit boards from the illustrations, then you know what to expect. Some experienced builders like to simply solder components to a solid unetched board in a technique referred to as *ground-plane* or *ugly construction.* If you find duplicating the etching illustrations tedious, much of the transceiver can be constructed in this way.

47

An overview

One way to describe this and other transceivers is to list the number of frequency conversions it performs before converting the IF signal to audio or the other way around. This design uses two intermediate frequencies. Using 72 MHz as the first IF allows some simplification when covering a large frequency span. Figure 4-1 shows how the major subassemblies are deployed.

The big advantage of using a high IF such as 72 MHz is that rejecting unwanted first-mixer products is possible with a single low-pass filter. With this scheme, the local oscillator and image are both far above 30 MHz. We don't have to use the multiple bandpass filters needed in designs with a much lower IF. I like to think of it as a 30-MHz-wide single-band scheme.

A lot of the blocks shown in figure 4-1 should look familiar. Scanning along the diagram reveals the CW-4.935-MHz oscillator to be the last previously discussed circuit. Everything else is new terri-

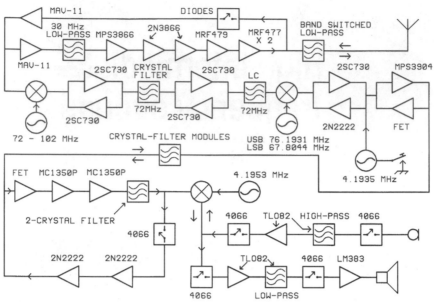

■ 4-1 *From microphone to antenna or antenna to loudspeaker, this is an overall view of signal processing in the transceiver. Transmitted CW signals start at the telegraph-key symbol. Power distribution and control circuits are not shown. The computer and synthesizer are major stories to be presented in upcoming chapters.*

tory. This chapter is concerned with the circuits leading up to the first mixer. The remaining RF circuits, except for the first local oscillator, are covered in chapter 5.

Some of the techniques used to build the RF power amplifier are a bit different from what we've covered so far, but they are not really difficult. Some broadband transformers will have to be constructed and most are even easier than the mixer transformers.

The first local oscillator, represented by the little generator symbol labeled in figure 4-1 as "72–102 MHz," is part of the frequency synthesis scheme. Much of the synthesizer is intimately involved with the three-chip computer, so there is much more to the first local oscillator issue.

As you have probably noticed, we have progressed in a matter-of-fact fashion, describing what is in the transceiver and how it works. I suppose it would be nice to include material about how calculations and design decisions were made, but the texts listed as references do an excellent job of that. I think it's important to document the radio circuits; we can then examine the computer and synthesizer principles more closely.

Bidirectional amplifiers

The next stage of the project is actually operating at the lower IF, but it connects directly to the 72-MHz mixer.

Two transistors share some common components, as shown in figure 4-2. The circled-A control line powers Q2 for transmitting. Transistor Q1 is powered by the circled-B control line for receiving. Negative feedback between collector and base is provided by R3 and C3 for either transistor. R5 is another shared component affecting the emitter current of both transistors.

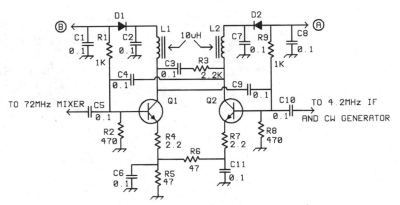

■ **4-2** *This bidirectional amplifier sends or receives signals to or from a mixer used in the 72-MHz IF strip. D1 and D2 are 1N4148 types or similar. Q1 is a 2SC730. Q2 is a 2N2222. Similar types of transistors could be substituted.*

This circuit could be changed to apply supply voltage to both transistor collectors continuously and control them with base bias currents. Unfortunately, I found that strong oscillations sometimes take place. It is easier to maintain stability by stopping collector current, along with base current, for the inactive transistor. Diodes D1 and D2 prevent current flow at the base-collector junction of the inactive transistor, which occurs from rectification of large signals.

For component locations, see figure 4-3. The pad between C8 and C4 is connected to a line which goes to another pad. This line represents a wire jumper which must be added to complete the circuit. Figure 4-4 shows the etching pattern for the layout given in figure 4-3.

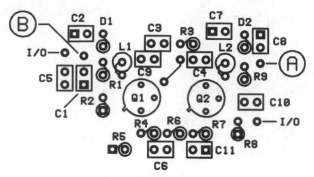

■ **4-3** *Layout for the lower-IF bidirectional amplifier.*

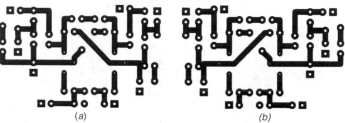

(a)   (b)

■ **4-4** *Etching pattern (actual size) for the lower-IF bidirectional amplifier: (a) for the photocopy toner method, and (b) for the photographic method.*

More on mixers

The mixer shown in figure 4-5 is almost the same as the one previously used as a product detector. All of the earlier construction tips apply because it uses similar parts and wiring. It does a reasonable job, especially for the cost involved. The port-to-port isolation and signal-handling capability may not be as good as those of some commercial mixers. Some of the suppliers listed in Appendix B have ready-made diode mixers.

Layout and lead length are more important for the mixer circuit at 72 MHz regarding balance and suppression of unwanted signals. If you design your own layout from the schematic diagram, keep this fact in mind. I didn't bother to match the diodes for forward resistance and capacitance; however, doing so may improve balance in some cases.

LC filters

After mixing with the second local oscillator signal, the lower IF signal is converted to approximately 72 MHz. A filter is needed to sup-

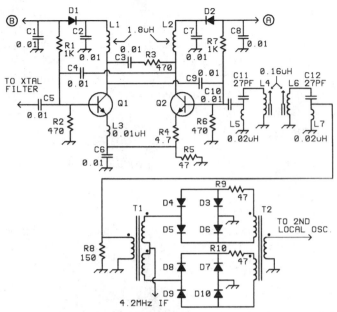

■ **4-5** *72-MHz mixer, filter, and amplifier. Parts are as follows: D1 through D10 are 1N4148 or similar. L5 and L7 are ½-in lengths of AWG number-22 wire. L4 and L6 Circuit Specialist Inc. part number 48A147MPC or Digi-Key part number TK2727-ND. Q1 and Q2 are 2SC730 or similar. T1 and T2 constructed from Amidon FT37-61 cores with five trifilar turns of AWG number-24 (or so) enamel-covered wire.*

press undesired mixing products. This is accomplished by an LC filter, as shown in figure 4-5, and later a crystal filter. The LC filter is made up of L7, L6, and C12 loosely coupled to L4, L5, and C11.

When tuning L4 and L6, some interaction occurs between the two tuned circuits. The final adjustments may require alternating between the two coil slugs. Be sure to use a nonmetallic tool to perform the alignment. Self-supporting air-core inductors will also work. They are harder to adjust because you must expand or compress the coil to tune them. The filter has a sharply peaked response, so tuning in this manner is not easy.

If the LC filter does not perform well, spurious emissions will occur when transmitting. The second local oscillator is only about 4.2 MHz away from the filter passband. The much stronger mixer image is 8.4 MHz above or below the filter passband. These far-removed mixer products are not attenuated sufficiently by the crystal filter alone.

Following the LC filter is an amplifier stage very similar to that shown in figure 4-2. Some component values are changed to be more appropriate for 72-MHz operation. Instead of a resistor in the emitter of Q1, an inductor is used for improved sensitivity when receiving. The exact value of L3 is nothing more than an estimate based on circuit paths, lead lengths, and C6.

Crystal filters

More close-in selectivity is provided by the crystal filter shown in figure 4-6. Y1 and Y2 are third-overtone quartz crystals ground for a series-resonant frequency of 72.0 MHz. The bandwidth of this filter is about 8 to 10 kHz wide, depending on the exact component values. I used small, self-supporting air-core coils for L11, L12, and L13. Final adjustments were made using the swept-frequency generator discussed in chapter 18.

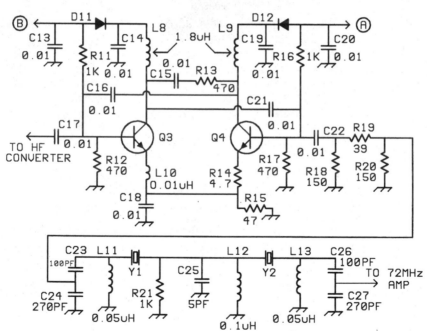

■ **4-6** *72-MHz crystal filter, attenuator, and amplifier. Parts are as follows: D11 and D12 are 1N4148 or similar. L11 and L13 are three turns of AWG number-22 wire wound on a ¼-20 bolt. Unscrew the bolt after the coil is formed. L12 is three turns of AWG number-22 wire wound in the same manner as L11 and L12. Q3 and Q4 are 2SC730. Y1 and Y2 are 72-MHz third-overtone crystals, Mouser stock number 332-1720.*

This is another place where you may want to use a commercially produced assembly, especially if you're short on test equipment. Some filters are available near 72 MHz and the synthesized local oscillator can be easily modified to accommodate other frequencies. Of course, even a ready-made filter will require some means of impedance matching unless it is designed for 50-Ω terminations.

The amplifier stage is the same as before and a resistive attenuator consisting of R18, R19, and R20 connects it to the crystal filter. The purpose of the attenuator is to provide a less reactive load for Q3, because Q3 is the load for the HF mixer. Input impedance for Q3 is therefore more constant because of the attenuator. The HF mixer termination thus sees a more constant impedance for various frequencies, which should preserve its strong-signal capability.

Assembling the high-IF board

Locations for the components of figures 4-5 and 4-6 are shown in figure 4-7. The LC double-tuned filter is surrounded by a rectangle, indicating a possible location for a shield container to prevent stray coupling to other circuit boards. I didn't find it necessary in the prototype. Enclosing the entire board would probably be a better idea.

You may have to reposition inductor L4 or L6 if you use coils physically much larger or smaller than called for in the parts list. If you must substitute, larger coils usually have a higher unloaded Q than smaller ones.

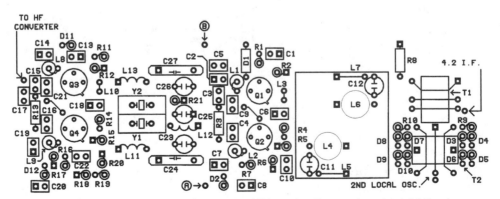

■ **4-7** *Component placement for the 72-MHz strip. Remember to add the jumpers near C9 and C16.*

Lines through T1 and T2 connected by doughnut-shaped pads represent an individual wire just as in the product detector layout discussed in chapter 2. Of course, the wire actually goes through the core several times instead of just once.

If you use similar but different transistors in the amplifiers, remember to find out if the collector is connected to the case. If so, place an insulating washer under the transistor to prevent contact with the ground plane. Place the transistor body snug against the washer for minimum lead length.

The coils used for L11, L12, and L13 are more or less stuffed in place. They can be tuned by spreading or compressing the turns. Etching patterns for the board layout are shown in figure 4-8.

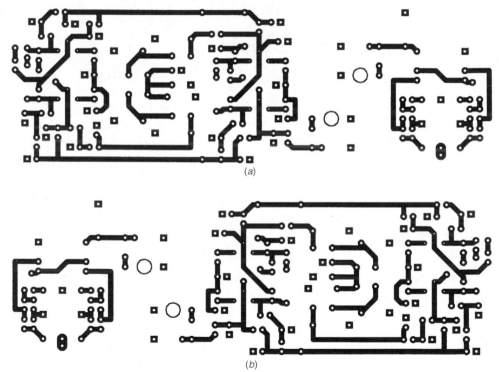

(a)

(b)

■ **4-8** *Etching pattern (actual size) for the 72-MHz IF strip: (a) for the photocopy toner method, and (b) for the photographic method.*

When time allows, I plan to try another crystal-filter circuit that should have less attenuation in the passband. See figure 4-9 for the alternative circuit. It should have less in-band attenuation and may be easier to adjust. At any rate, there are fewer components. If you decide to try it, place parts as indicated in the layout shown in figure 4-7.

4-9 *An alternative circuit for the 72-MHz crystal filter.*

Second local oscillators

Two oscillator circuits generate the signal needed by the mixer to generate new frequencies. One runs above the 72-MHz IF and the other one below it. The reason for this is that the lower-IF crystal filter has an asymmetrical response. It works well as a lower sideband filter, but a method was needed to produce upper sideband signals as well.

When using the 67.8044-MHz oscillator, the lower sideband signal is translated to 72 MHz as is. The lower IF and second local oscillator frequencies add to form the new frequency. When the 76.1931 oscillator is used, the signal at 72 MHz is inverted to an upper sideband signal. In this second case, the difference is used. For instance, if the frequency of the lower IF signal rises slightly, the output frequency moves down by the same amount.

An overtone-type crystal oscillator is used to generate a reasonably stable signal at these frequencies. The oscillator circuits appear in figure 4-10 along with amplifier stages to increase the output level. Third- or fifth-overtone crystals will work here. Oscillation takes place at the overtone frequency, even though the crystals are capable of oscillating at their fundamental frequency. This is not a frequency multiplier circuit where harmonics of the fundamental frequency are used.

Oscillator components around Q1 and Q2 force oscillations to take place near the correct frequency. For instance, L2, C5, and C6 form a tuned circuit. You can replace Y1 with a low-value resistor and the circuit will again oscillate—not that you would want to do this, because stability would be much poorer without the crystal.

A control circuit, to be described later, supplies power to one oscillator or the other. For the Q1 oscillator, diode D1 is forward-biased and permits signal flow from the link winding of L2 to IC1. The other diode, D2, with no forward bias, helps isolate the Q2-tuned circuit and oscillator. When the Q2 oscillator is active, the Q1 oscillator is isolated in the same manner.

Amplifier stages IC1 and IC2 are monolithic integrated circuits with several built-in features which make them easy to use. They are designed for use in systems with 50-Ω input and output imped-

55

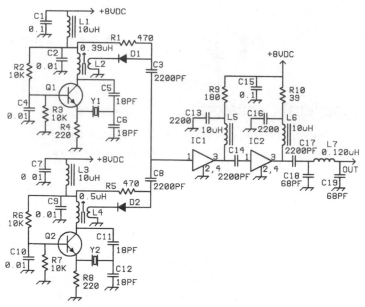

■ **4-10** *Second local oscillator and amplifier circuits. Parts are as follows: D1 and D2 are 1N4148 or similar. IC1 is a Mini-Circuits MAR-1. IC2 is a Mini-Circuits MAV-11. L2 is a Mouser part number 434-1012-8.5C or similar. The link winding consists of ¾ of a turn. L4 is a Mouser part number 434-0712-10.5C or similar. The link winding consists of ¾ of a turn. L7 is made of six turns of AWG number-22 wire wound on a ¼-20 bolt. Unscrew the bolt after the coil is formed. Q1 and Q2 are MPS3866 or similar. Y1 is a third- or fifth-overtone crystal ground for series resonance at 76.1931 MHz. Y2 is a third- or fifth-overtone crystal ground for series resonance at 67.8044 MHz.*

ances. When used correctly, they are unconditionally stable. Most of the components for biasing are already built in. Discrete transistor stages could certainly be used here, but the ICs take up very little space. Resistors R9 and R10 set the dc operating points for these ICs.

After being raised to approximately 50 mW, the second oscillator signal from IC2 goes through a low-pass filter made of C18, L7, and C19 on the way to the mixer. The filter will probably work well enough on other types of mixers without changes, even though they may present a load closer to 50 Ω than the 125-Ω port of the dual-bridge mixer. However, make sure it is rated for the oscillator injection power level.

Second oscillator containment

When assembled on a board, the oscillator and amplifier resemble figure 4-11. For the etching pattern, see figure 4-12. Looks simple enough, doesn't it? Maybe, but for me it turned out to be a real troublemaker. For one thing, signals around 70 MHz are easily radiated into other parts of the radio. Another reason is because of the output power, even though it helps boost strong signal capabilities of the diode mixer.

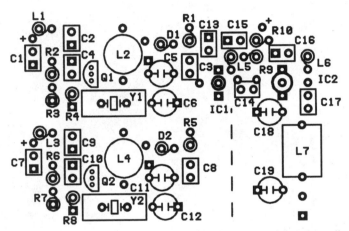

■ 4-11 *Component placement for the second local oscillator board. The MAR-1 chip has a dot near pin number 1. The MAV-11 has a dot near pin number 3.*

57

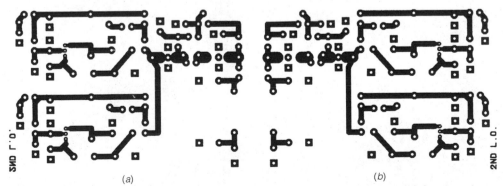

(a)   (b)

■ 4-12 *Etching pattern (actual size) for the second local oscillator: (a) for the photocopy toner method, and (b) for the photographic method.*

Good shielding for this circuit board is a must. It is also a good idea to use feed-through capacitors for routing dc power into the enclosure. Of course, the output connection should be wired with miniature coaxial cable. Solder the cable shield to the enclosure wall, not to the board ground plane. Alternatively, mount a coaxial connector and plug.

Problems caused by signal radiation or leakage from this board were related to reception. The more severe the problem, the more "birdies" or heterodynes you will hear, and they will be louder. There are plenty of signals on the bands to listen to without adding artificial ones, so shield the board well.

Fundamental frequency and control circuits

5

AT LAST, WE REACH A POINT WHERE THE SIGNALS GENERated by the circuits described in preceding chapters can be converted to the actual operating frequency and sent to the antenna. Conversely, faint signals can be selected from almost 0 to 30 MHz by reversing the process for reception.

Many of the receiver circuits are similar to previous examples and use many of the same types of components. Some of the transmitter circuits handle signals of a vastly larger power level. The power amplifier components are larger but no more difficult to work with.

A stable continuous RF source is needed to finish the frequency conversion. This signal is provided by indirect synthesis and will be described in later chapters. An external signal generator can be used as a substitute when testing.

High-frequency conversion

The oscillator signal needed by the mixer is generated by a voltage-controlled oscillator in the frequency synthesizer. In figure 5-1, capacitor C13 couples the VCO signal to Q1. The VCO signal is amplified by Q1, filtered, and sent to T2.

IC2 amplifies signals from the mixer when transmitting. IC1 is used as a preamplifier to boost signals coming from the antenna. Of course, signals actually must go through a filter and the transmit/receive (T/R) switch before reaching IC1. Diodes D1 and D2 are used to gain some isolation when an IC is unpowered, by virtue of the characteristic junction potential with no forward bias.

Some operating conditions, such as transmitting into a severe impedance mismatch, can overwhelm the attenuation capabilities of the T/R switch. Diodes D11 and D12 are used to protect IC1

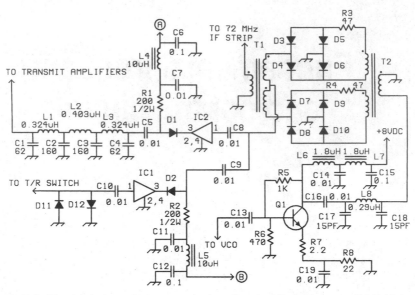

■ 5-1 *The high-frequency conversion circuit. All diodes are 1N4148 or similar. IC1 and IC2 are Mini-Circuits MAV-11 types. L1 and L3 consist of 10 turns of number-22 wire wound on a ⅜-in-diameter bolt with 18-turn-per-inch threads. Remove the bolt or use a nylon commode-bolt for a permanent form. L2 uses 11.5 turns and L8 uses 9 turns on the same form. Q1 is an MPS3866 transistor. T1 and T2 are wound with 5 trifilar turns of number-26 or smaller wire on Amidon FT37-43 cores.*

from being damaged by such actions. A useful addition at the input of IC1 would be some sort of manually adjustable attenuator. Such attenuation is useful when the receiver is exposed to extremely strong signals, which can cause false responses or blocking of desired signals.

R1 and R2 serve a dual purpose. They set the dc operating point for IC1 and IC2. They also allow each IC to develop an output signal at very low frequencies (VLF) because they do not change impedance with frequency as an inductor would.

Using the center-tap connection of T1 as a mixer port also allows the mixer to work at VLF. I haven't really explored the VLF receiving capabilities of the transceiver other than to verify that it works. If you plan to use this frequency range extensively, capacitors C8, C9, C10, and C11 could be changed to 0.1 µF for better weak-signal reception.

Figure 5-2 is a placement diagram for parts on the HF converter board. Inductors L1, L2, and L3 are all mounted at right angles to each other. L1 and L3 are oriented in the same plane. The axis of L2 is pointing up, out of the board. The coils should not be mounted too close to the board ground plane. Try to keep L1 and L3 at least a half diameter away from the board. L2 should be at least one diameter above the board. I glued the coil forms to small pieces of Plexiglas with a cyanoacrylate glue. The Plexiglas supports are also glued to the circuit board.

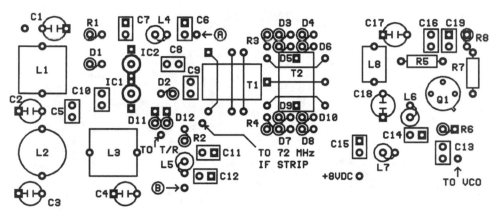

■ **5-2** *Locations of parts on the HF converter. For clarity, lines through T1 and T2 show the path of each individual wire in the winding as though it makes only one pass through the core.*

Prior to mounting them, drill a hole slightly larger than the body of each IC. Don't forget to add jumpers from each square pad to the ground plane. You may want to drill small holes alongside the place where the IC ground leads lie instead of directly in-line. These steps will allow all four leads to seat flat against the board. Patterns for etching the board are shown in figure 5-3.

Building transmitter power

Coming out of the low-pass filter shown in figure 5-1 is a small version of the signal which you are going to put on the air. All that's left to do is amplify it to a suitable level with as little distortion as possible. Figure 5-4 presents a three-transistor circuit capable of bringing the signal up to about 300 mW. It includes a remotely controlled attenuator for adjusting the input level.

Drive control R3 controls the conduction of D1, allowing it to shunt more or less of the input signal to ground as needed. As shown, R3 allows an adjustment range of 10 to 15 dB. On most bands, the prototype transceiver has a minimum output power of

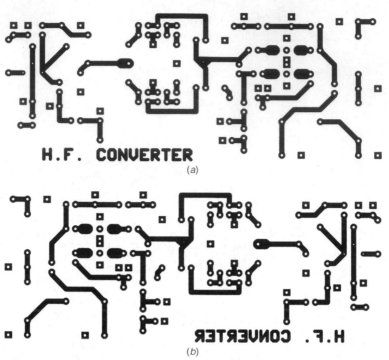

■ **5-3** *Etching patterns (actual size): use (a) for photographic reproduction and (b) for the transparency iron-on method.*

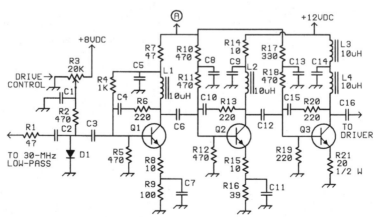

■ **5-4** *The transmitter preamplifier chain. All capacitors in this circuit are 0.1-μF monolithic ceramic types. Diode D1 is a 1N4007. L1 through L4 can be small molded chokes or toroidal inductors. Q1 is an MPS3866. Q2 and Q3 are 2N3866 types. Resistors are ¼ W unless labeled otherwise.*

about 5 W. For those who like to try milliwatt operating, it is more efficient to modify the bias of the driver and final RF stages than to modify the attenuator. This will be covered shortly.

Two of the transistors have enough standing current to require a heat sink. For Q2, a small push-on "hat" is sufficient. Q3 requires more heat removal. Q3 should be clamped to a heat sink with a thermal resistance of approximately 20°C/W or less. In the prototype I used a small clamp-on heat sink, which I bolted to a larger, finned heat sink, using insulating hardware and silicone grease.

Circled A is again the control line which enables this circuit during transmissions. Both input and output connections should be made with miniature 50-Ω coaxial cable. Parts are located as shown in figure 5-5. The transistors should be mounted close to the board with insulating washers underneath Q2 and Q3. See figure 5-6 for the etching patterns.

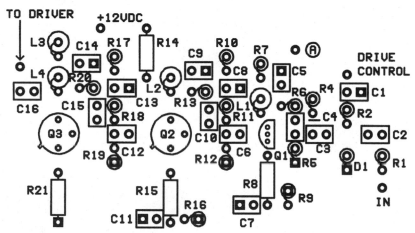

■ **5-5** *When installing heat sinks on the transmitter preamplifier chain, take care to avoid shorting to other components.*

As mentioned in chapter 2, this board is part of the amplifier chain necessary for amplifying the weak double-sideband signal before sending it to the antenna. The driver, power amplifier, and part of the filter board can be used to finish out the double-sideband system. So, whether you've skipped ahead to this point to make the DSB transceiver or plan on building the larger multiband transceiver, let's proceed to the next transmitter stage.

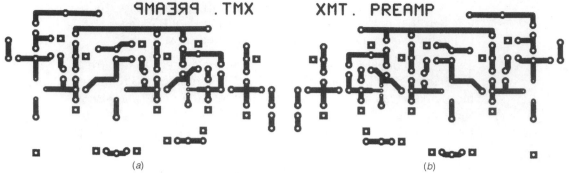

(a)

(b)

■ **5-6** *Etching patterns (actual size): use (a) for the transparency iron-on method and (b) for the photographic reproduction method.*

Driver and power amplifier

As we build amplifying stages capable of handling larger signals, practical considerations require doing things a bit differently than we did with the previous circuits. Figure 5-7 shows Q1 being used as a driver stage for the final amplifier. The final is a push-pull design. I will attempt to explain why these amplifiers look different from their low-powered counterparts.

If we increase the power capability without raising the supply voltage, that means larger currents and lower impedances. The easiest way to match said impedances and transfer power over the HF spectrum is with wide-bandwidth transformers. This is the reason for transformer coupling between stages.

When the driver stage is in operation, T1 passes the input signal to Q1. T1 also matches the low impedance present at the base of Q1. Any capacitance connected to the windings of T1, along with its own parasitic capacitance, can resonate and encourage oscillations. Resistor R2 is a swamping resistor for lowering the Q of T1 to discourage such occurrences. Negative RF feedback is provided by R1 and C2. This shunt feedback also aids stability and helps equalize the frequency response of the driver stage.

Transformer T2 matches the driver stage output to the final amplifier input, at least more closely perhaps. Matching is only approximate because transistor impedances vary over the wide frequency range covered by these stages. Capacitors C5 and C6, along with resistors R3 through R6, help equalize gain over the frequency range used for transmitting.

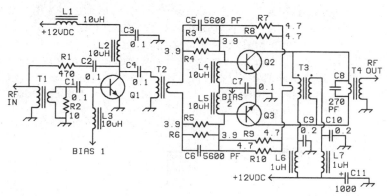

■ 5-7 *Pump up your signal with the driver and final amplifier board! L1 and L2 must have a current rating of 1 A or greater. Three turns of number-22 wire on an Amidon FB43-801 core will work. Q1 is a Motorola MRF479. Q2 and Q3 are Motorola matched-pair MRF477 transistors. Transformers T1 and T2 are made with Amidon FB43-6301 ferrite beads and number-22 enameled wire. T3 uses an FT50-61 toroid and number-22 enameled wire. The core for T4 is an Amidon BN43-7051 and uses number-24 plastic-covered hookup wire.*

Negative RF feedback is also used in the final amplifier. Instead of connecting directly to the collectors, a single-pass conductor through T3 sends out-of-phase current through resistors R7, R8, R9, and R10. T3 also serves as an artificial RF center tap for T4, while providing separate dc paths for Q2 and Q3. The separate paths are also an aid in stabilizing the amplifier. T4 completes the output circuit by providing an impedance transformation.

Many of the ideas for the final amplifier came from Motorola applications literature. The output stage has been operated into severe mismatches at full power, both intentionally and by accident, with no damage. It is not unconditionally stable, but as long as the VSWR is below about 1.7, no oscillations should take place.

Many low-powered stages in previous circuits used emitter resistors to provide negative feedback and assure stable dc operation under varying thermal conditions. Doing so at higher power levels is usually not practical because of RF instability. In the interest of avoiding parasitic oscillations, the emitters are connected directly to ground. Thermally compensated bias for the driver and final is supplied by external circuitry.

Bias circuits for both stages are shown in figure 5-8. D2 and D4 are really the diode junctions of two tab-mounted NPN transistors. The emitter and collector are used as a cathode. The tab is bolted directly to the heat sink, thereby providing an electrical and thermal connection. The base lead is used as the anode for D2 and D4. Before mounting these transistors, solder the collector and emitter lead together.

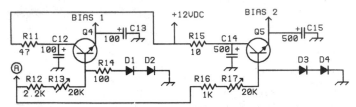

■ **5-8** *Base bias for the driver and final amplifier is supplied by this circuit. C13 and C15 voltage rating should be around 3 to 6 V. C12 and C14 should be 16 V or more. D1 and D3 are 1N4148 silicon diodes. D2 and D4 are TIP29, NPN power transistors (see text). Q4 is a 2N2222A or similar. Q5 is a TIP47 or similar.*

As the driver and final transistors rise in temperature, so do the sensing diodes bolted to the heat sink. If, for instance, D2 is heated, it will draw more current and decrease the output of emitter follower Q4. This action protects the driver stage from thermal runaway, which could occur as Q1 draws more current when its junction temperature rises.

Bias currents for driver and final stages are turned on for transmitting and off during reception by the circled-A control line. This helps reduce power drain in receive mode, but it also stops the amplifier chain from acting as a wideband noise generator. Unless a relay or T/R switch with very high attenuation is used, some of this noise may be picked up during reception.

For best results, D2 should be mounted close to Q1 and D4 should be mounted close to Q2 and Q3. The prototype transceiver has a less-than-optimum mounting position for the sensing diodes because I used space on the amplifier board to mount the sensing diodes and bias circuits. I also used a large heat sink instead of two separate ones for driver and final stages. Despite these compromises, the bias circuits do an adequate job of regulating standing currents for both stages. Layouts described in this chapter allow more freedom in placing the sensing diodes.

Potentiometer R13 is used to adjust the standing or quiescent current in Q1, and R17 serves the same purpose for Q2 and Q3. I normally set the standing collector current of Q1 to about 1 A for class A operation. I usually set the standing current for the final stage to about 3 A. This is much higher than required for class AB1 operation and is not really necessary. Doing so seems to improve the linearity of the final stage, but it should not be set so high unless you have a sufficiently massive heat sink and/or fan for adequate cooling.

A useful trick for those who want to use the transceiver occasionally for very low power CW operation is to install a switch in series with the circled-A control line. With no bias to either stage, gain is reduced and power levels well below 1 W can be achieved. Two additional benefits result from this modification. Power consumption during transmit is reduced because of more efficient class C operation. You have, in effect, two ranges of adjustment, making the drive control easier to use at the lowest output levels.

Amplifier components

With very low impedances present in many of the RF paths, it is important that resistors have very little parasitic inductance. Use carbon-composition resistors for the driver and final amplifier circuits. Most common carbon-film and metal-film resistors have a spirally shaped element with too much inductance, as do wire-wound resistors.

Capacitors must be low-inductance types. I used small monolithic ceramic capacitors with wire leads. Chip-type capacitors designed for surface-mounting can be used by soldering them to the bottom side of the board.

Inductors are also special in some instances. L3, L4, and L5 can be just about any kind of small RF choke. The other inductors must be capable of handling larger currents, especially L6 and L7.

Transformers T1, T2, and T4 are called *conventional* transformers because they don't use transmission line elements. Yes, we have to construct them, but they're even simpler in some ways than the previously discussed transformers. For all three units, the low impedance winding is a single pass of a large conductor through the ferrite material.

For T1 and T2, I used copper braid stripped from RG-59 coaxial cable as the low-impedance winding. Figure 5-9 shows how it is threaded through two ferrite beads. Don't flatten the braid. Leave it in a round shape as you place it in the beads. You can use a drawing compass, test probe, or other pointed object to keep the braid inside open while you flatten the ends and crossover section between beads. The idea is to leave it in a cylindrical form inside the beads.

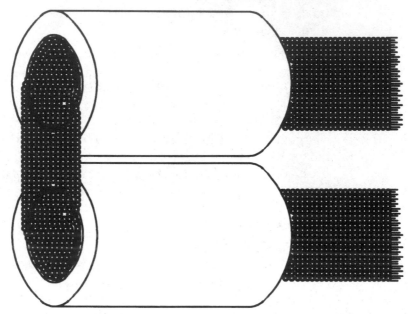

■ **5-9** *Make transformers T1 and T2 by placing two ferrite beads side by side. See text for primary and secondary winding instructions.*

For T1, ends of the braid are later connected to ground and C1. The primary winding is formed by passing the end of an insulated wire through the small tunnel formed by the copper braid inside the bead. Bend the wire around and pass the end through the other bead going in the opposite direction. Bend it again and thread it back into the first bead inside the braid tunnel. Keep doing this for T1 until you've completed four turns. In other words, there should be four short lengths of wire connecting the beads on the end where the braid hangs free.

Each bead should have braid extending from one end and a wire extending from the other end when the transformer is completed. For T1, one wire end connects to ground and the other connects to the RF input terminal. Transformer T2 is fabricated similarly, except it uses only two turns of wire. Figure 5-10 shows how the wire would look if the beads and braid disappear. Neither end of the braid on T2 is grounded.

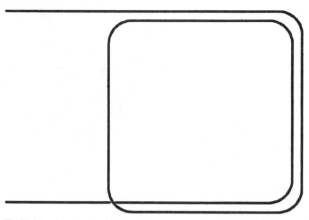

■ **5-10** *A view of how the enameled wire is wound through T2.*

Transformer T4 uses a solid block of ferrite material with two parallel holes in it: a binocular core. It can be constructed in the same manner as T1 and T2 or as I did with the prototype. Instead of copper braid, I used thin-wall brass tubing. The two lengths of tubing are cut slightly longer than the core. Two pieces of single-sided circuit-board laminate are drilled to accommodate the tube ends. Each piece of laminate is placed over the ends of the ferrite core, with brass tube ends slightly protruding through the holes.

To complete the low-impedance winding, I solder the brass tubes to the foil and cut the foil midway between the holes on one of the board-laminate pieces. Each side of the cut is one end of the primary winding. Four turns of plastic-insulated AWG number-26 wire are used for the secondary winding in this transformer.

Transformer T3 is a transmission line type similar to the smaller ones used previously. The major difference with T3 is that it uses more conductors. This lowers the transmission line impedance and reduces dc voltage drop across the windings. Six wires are twisted together before being wound through the core. As before, the pitch is not particularly critical. You should probably avoid twisting them tighter than two turns per inch because winding the twisted bundle through the ferrite core will become difficult. Make five passes through the core of T3.

Locations for all of the driver and final components are displayed in figure 5-11. A single-turn link connecting R8 and R9 is represented by a broken line through the core of T3. The other two multiturn windings are depicted as solid lines which cross in the center. Each solid line represents three wires, or, to put it another way, half of the transmission line section.

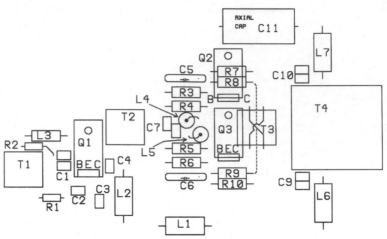

■ 5-11 *Part locations for the driver/final board. Although not apparent in this drawing, power transistors Q1 through Q3 are actually mounted underneath the board and bolted to a heat sink.*

To connect T3, all you have to do is separate three wires from one end of the bundle or cable after stripping all six wire ends. Next, twist the three ends together. Use an ohmmeter or markings to find the other three ends, separate them, and twist them together to form a new, smaller bundle. This takes care of one winding. The remaining three wires form the other winding. Connect the windings as shown by the solid lines drawn through T3.

Most of the components are mounted in a conventional manner except the sensing diodes, transistors, and some transformers. The transistors are mounted on the underneath or etched side of the board. This is because their mounting tabs must be bolted to the heat sink. Leads are bent at a right angle, inserted into the board, and soldered. The best way to position the transistors is to wait until you have the board bolted in place along with the unsoldered transistors stuffed into the board. Tighten mounting screws into the transistor tabs to make sure everything is lined up correctly without excessive strain or binding on any of the transistor leads.

If everything fits well, including the sensing diodes, solder the center emitter lead of each transistor to the ground plane. Carefully unbolt everything, taking care not to disturb the positioning of the transistors. Solder base and collector leads of each transistor to the foil on the backside of the board. You probably should mount the transistors last when building the board.

Before remounting the board for the final time, make sure no raised areas or irregularities exist where transistor tabs contact the heat sink. Drilling or tapping operations can leave a slight ridge around the hole. This must be cut away or countersunk to allow the tab to fit absolutely flat against the heat sink. Don't leave dirt, cuttings, flux, solder, or any other contaminant on the heat sink or transistor tabs. Dab silicone heat-sink grease on each transistor tab. These precautions also apply to the sensing diodes.

The heat sink on the prototype measures 4.5 in wide, 1.25 in thick, and 7 in long. The base of the heat sink supporting the fins is 0.305 ($\frac{5}{16}$) in thick. A suitable commercial extruded aluminum heat sink would be Aavid Engineering 60140. If you can't find a single large heat sink, one may be fabricated by bolting smaller heat sinks to ¼-in aluminum plate. If possible, try to arrange the fins vertically.

Etching patterns for the driver and final stages are displayed in figure 5-12. Solder a number-12 or larger insulated wire to the center of the wide path spanning the output end of the board. This will be used for the power connection. It's all right to drill mounting holes through this path as long as precautions are taken to make sure it doesn't short to ground through the mounting hardware. Countersink the hole and use nonconductive mounting hardware as necessary.

The funny thing about transformers T1, T2, and T4 is the fact that they are surface-mounted on top of the board. Figure 5-13 is the etching pattern for the top side of the driver/final circuit board. So now the truth is out. This is one of those awful double-sided boards. Well, maybe not that awful. Both sides have large, simple patterns.

Registration, assuring proper alignment of one layer in relation to the other, is usually the biggest problem for do-it-yourself construction. One method of solving the problem is to mask and etch the board in two separate steps, as if it were two separate boards. Etch, wash, and dry the board in the normal manner as if you were going to complete only the bottom layer. So far, this would be the same procedure used to fabricate all of the previously discussed boards.

Next, pick three or four widely separated places on the board where a small hole coincides in both layers. This would be a place where a lead is used to ground a leg of T1, for example. Carefully drill a hole in these locations. They will be used to position the

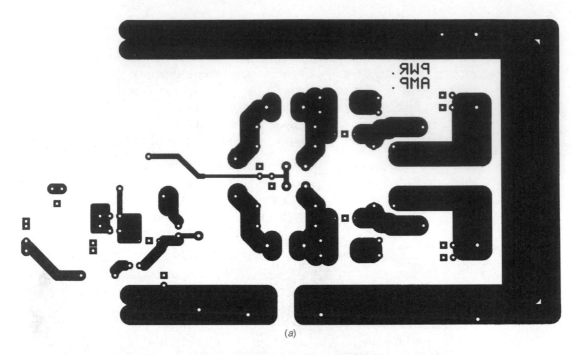

(a)

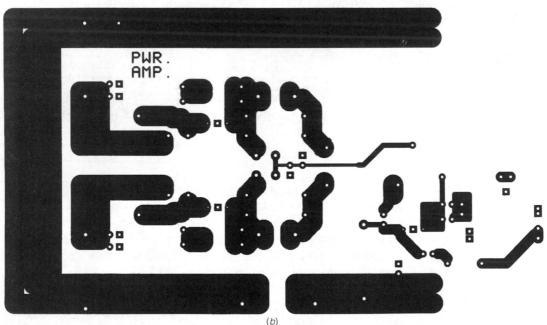

(b)

■ **5-12** *Etching patterns (actual size) for the bottom side of the driver/final board. The pattern at (a) is for transparency iron-on reproduction, and (b) is for the photographic method.*

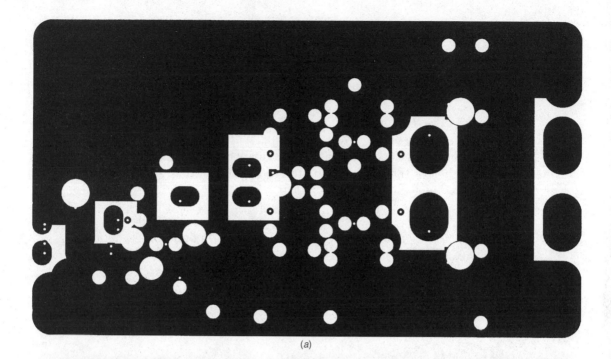

(a)

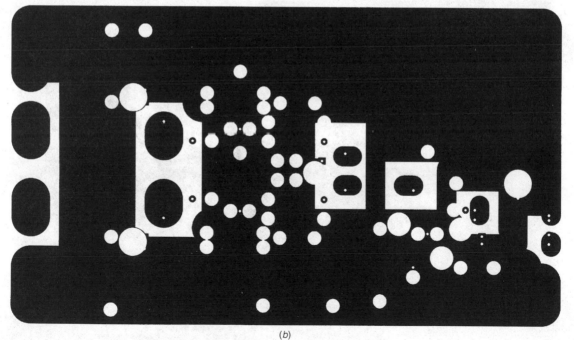

(b)

■ **5-13** *The top side of the driver/final board (actual size). If you overlay a copy of figure 5-11 on (a), the transistor tab holes will align with holes in the (a) pattern which must be drilled. This will allow you to access the screw heads with a screwdriver blade. The pattern at (a) is for transparency iron-on reproduction, and (b) is for the photographic method.*

iron-on or photo pattern. Paint or mask the bottom layer and bare the top layer, removing paint, tape, or whatever you used to mask it. Drill similar holes in the appropriate places on the pattern. Use pins, small drill bits, or wire leads to guide the pattern into place.

Secure the top pattern to the board with heat or tape, depending on which method you are using. With the top pattern fixed in place, pads and holes on the top side will be accurately positioned. Transfer the pattern and etch the board again, and it's ready for cleaning and drilling.

Another possible way to handle the top layer would be to leave a solid, unetched, copper ground plane and countersink all necessary holes. Transformer connections could be accommodated by gluing small pieces of board laminate on top for pads. Although it isn't exactly the same, the completed driver and amplifier board will look much like that in figure 5-14.

■ **5-14** *The layout for the prototype driver/final board looks slightly different because of the bias circuitry; however, the RF circuitry is the same.*

Bias regulator board

The bias circuits are not sensitive to board layout and physical configuration like the previous driver/final board. The driver/final board is not easy to rework or experiment with, because it must be bolted to the heat sink. For these and other reasons, it makes sense to construct a separate board for bias circuits. It is small enough to mount on top of the unused area of the larger board.

The sensing diodes shown at the top of figure 5-15 mount on the heat sink, as mentioned earlier. This board can be constructed by etching the bottom side, drilling holes, and stuffing components, as with all of the previous boards. However, another approach might be more appealing. With conventional construction, solder connections are underneath, requiring the board to be mounted on spacers. You have no access to solder connections with it piggybacked on the larger board.

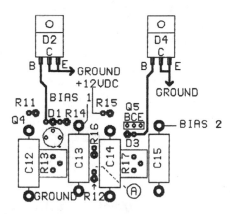

■ **5-15** *Parts layout for the bias board. D2 and D4 are not mounted on this board but are shown to indicate how they are connected.*

Consider surface-mounting. Why not use the mirror image of what you've been doing and print the pattern on top? Mount components by simply bending leads to form a small foot and solder them to the pads. To increase the pad area, color in all the pad holes with a permanent marker before etching. Use good-quality board stock because there are no holes to help support the component leads. Another trick for strengthening this kind of construction is to glue the bodies of the larger components to the board.

Now you have access to all points of the bias circuit, and the board can be mounted flat against the larger board with double-backed tape, glue, or screws. This is an interesting way to make boards, and I've built entire projects this way. One amateur I know personally, Don Kelly, KA5UOS, built dozens of projects using this method. Figure 5-16 shows both views of the etching pattern, so you can construct the board with the method or layer of your choice.

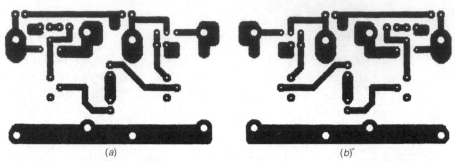

(a)   (b)"

■ **5-16** *The pattern at (a) is for transparency iron-on reproduction and (b) is for the photographic method (actual size). See text for suggestions in using surface-mount techniques.*

Filtering

If the amplifier chain used for transmitting had no harmonic distortion, we could simply connect it to the antenna terminal. This is not the case; therefore, low-pass filters are used to bring harmonic emissions to acceptable levels. Additional circuitry selects and switches the proper filter in or out as commanded by the 8031 microcontroller.

Filter selection commands come to the filter board in the form of parallel 4-bit binary signals. They are sent out by a 4094 shift register not located on this board. The 4094 in figure 5-17 is powered by 5 V, while IC1 is powered by the 12-V supply. Translation of 5-V logic levels to 12-V levels is accomplished by Q1 through Q4.

IC1 is a 1-to-16 multiplexer/demultiplexer. Binary signals on its selector pins "connect" the common pin to one of 16 input/output pins. The common pin is wired to the 12-V supply, allowing pin 8 to supply power to relays K1 and K2, which connect the 160-m filter. If pin 7 is selected, the 80-m filter is engaged, and so on.

Figure 5-18 gives component values for all of the low-pass filters. The capacitors are standard values; inductors are not. Inductors must be fabricated to meet the values indicated. Most of the inductors can be wound on powdered iron cores with a specified number of turns and result in inductance values close enough for proper operation. I measured the inductance of all of the inductors

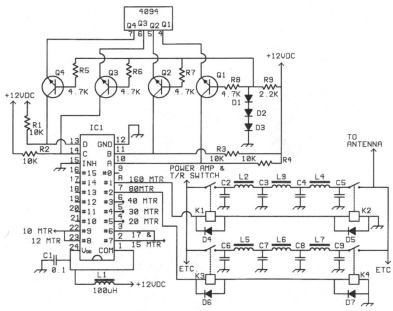

■ **5-17** *Partial schematic of the filter board. Diodes are 1N4148 or similar types. IC1 is a CD4067. Almost any small-signal silicon NPN transistor such as the 2N2222 can be used for Q1 through Q4.*

in the prototype when fabricating them. Measuring them with an inductance meter is the easy way.

You can also make measurements in an indirect way with a calibrated dipmeter or signal generator. Using an accurate fixed capacitance, calculate the resonant frequency.

$$\text{Frequency} = \frac{1}{2\pi\sqrt{LC}} \qquad (5\text{-}1)$$

and adjust the inductor to resonate the tuned circuit. In the case of a dipmeter, take the time to become familiar with its operation. Even with measurements it's best to make sure you know roughly how many turns are needed so that you're not fooled by a parasitic resonance or other false indication.

To use an RF generator, you will also need some kind of detector such as an RF voltmeter or scope in parallel with a low-value resistance of 50 Ω or so. With such a system, connect generator and detector grounds together. Use the parallel-tuned circuit to connect generator output to detector input. A sharp dip in the detector reading will occur as the generator is tuned across the resonant frequency.

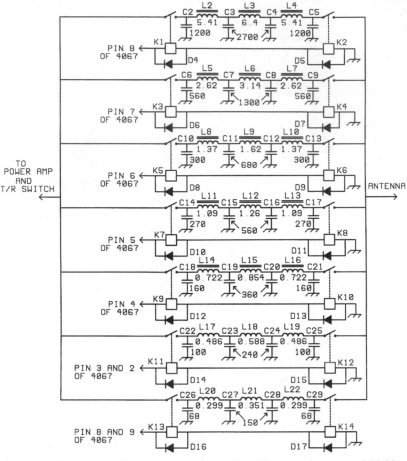

■ **5-18** *More of the filter board schematic. All capacitors are 200 V or greater and 10 percent tolerance or better. Capacitance values are in picofarads. Diodes are 1N4148 or similar types. All relays (K1 through K14) are 5-V reed relays, Radio Shack catalog number 275-232 or Mouser stock number 431-1405. Inductance values are in microhenrys. Evenly spaced turns to fill an Amidon T68-2 toroid: L2 and L4—30, L3— 33, L5 and L7—21, L6—23, L8 and L10—15, L9—16, L11 and L13—13, L12—14, L14 and L16—11, L15—12. Solenoid inductor dimensions for L17 through L21 are for number-22 wire wound on a ⅜-in-diameter bolt with a thread pitch of 13 turns per inch. Remove the bolt or use a nylon commode-bolt for a permanent form. Number of turns needed: L17 and L19—13.5, L18—16, L20 and L22—9, L21—10.*

The transmit/receive switch

An electronic T/R switch also resides on the filter board. The circuit shown in figure 5-19 is partially controlled by the circled-A control line. Diodes D20 through D23 are forward-biased by current from Q5 when receiving.

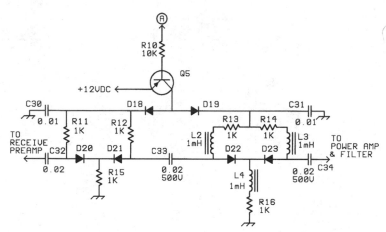

■ **5-19** *This circuit is an electronic transmit/receive switch which opens or closes the path between the antenna and receiver preamplifier. Diodes D18 through D21 are 1N4148 types. D22 and D23 are 1N4007 types. Q1 is an MPS2907 or almost any small silicon PNP transistor with a collector-to-base breakdown voltage of 60 V or greater.*

During transmissions, Q5 is turned off by a high on the circled-A control line. The large RF transmitter signal is rectified by diodes D20 through D23. The rectified current can't pass through Q5, so dc potential generated by the diodes reverse-biases them. With a strong reverse-bias voltage, they attenuate the signal before it reaches the receive preamplifier. Use miniature 50-Ω coaxial cable for RF connections to the receive preamplifier and power amplifier.

Locations for most components on the filter board are shown in figure 5-20. Five filters for the lowest frequency ranges use inductors made from iron-powder toroids. The two highest ranges use solenoidal coils.

Diodes D4 through D17 are not shown. If used, they must be soldered across each relay coil on the bottom side of the board. They reduce back EMF induced in the relay coils when the current is

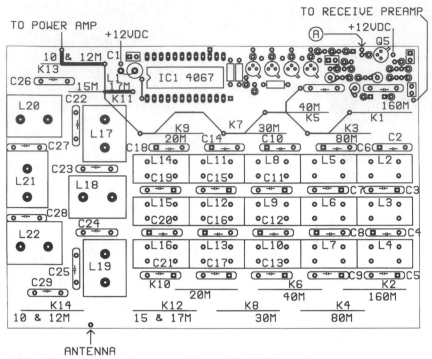

■ **5-20** *Locations for some of the larger components on the filter board. Lines labeled K1, K2, K3, etc., represent reed relays.*

stopped by IC1. The board currently is operating without these protective diodes with no apparent ill effects. It's a good idea to install them, because even if IC1 is tough enough to withstand the back EMF, it may cause effects, such as false triggering, in other circuits.

The heavy line which meanders along the odd-numbered relays (K1, K3, etc.) consists of point-to-point wiring. Use insulated wire and make sure none of the connections bridge to another conductor. For a close-in view of the smaller components associated with the T/R switch, see figure 5-21.

Level-shifting transistors are located alongside the T/R-switch components. Instead of connecting directly to the 4094 shift register, a better idea would be to use a filter mounted on the wall of the enclosure to keep noise from the digital circuits away from the receiver input. Figure 5-22 is the etching pattern for the transparency method and figure 5-23 is the pattern for the photographic method of transferring the image to a circuit board.

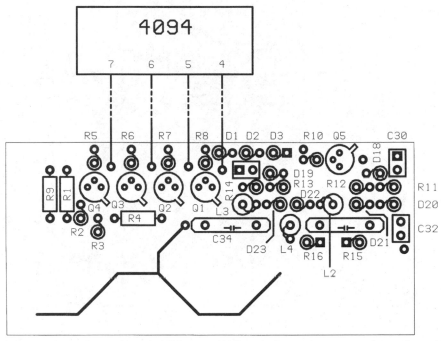

■ **5-21** *Locations of level shifting and T/R switch components on a portion of the filter board.*

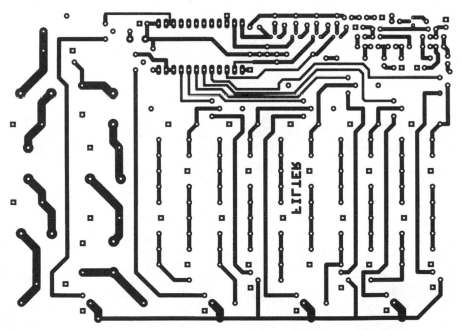

■ **5-22** *Filter board etching pattern for use with the iron-on trasparency process. Remember to set photocopier for 125 percent enlargement when making the transparency.*

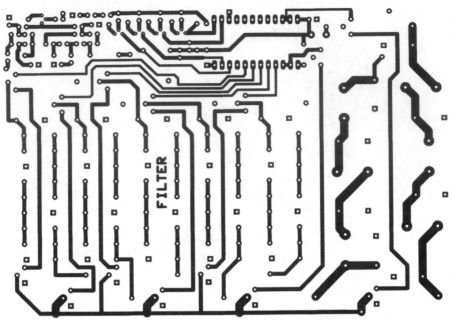

FILTER

■ **5-23** *Filter board etching pattern for use with the photographic process. This pattern must first be reproduced full size by enlarging it 125 percent to make contact (/:/) exposures.*

Tying it all together

Except for the controller and synthesizer, all of the major building blocks of the synthesizer have been described at this point. If you have been testing boards and assemblies, it is obvious that a lot of different points must be controlled to put everything in operation. The purpose of the control board shown in figure 5-24 is to automate many of the transceiver functions.

The most often described control functions up to this point have been the lines used for switching from receive to transmit operation and back. Although a toggle switch could be used, the circuit made of metal-oxide semiconductor field-effect transistors (MOS-FETs), Q1 through Q4, is more versatile. When the microphone button and key are open, Q1 and Q3 keep the circled-A control line at ground potential. Besides having practically no voltage, it will sink a considerable amount of current. It provides a low-resistance path to ground for controlled circuits.

When a microphone switch or key is closed, the circled-A control line rises to almost the same potential as the supply line. It can supply current and power to various circuits. Transistors Q2 and Q4 are driven by Q1 and Q3. Being an inverter, Q2 and Q4 set the circled-B control line to the opposite state of whatever the circled-A is.

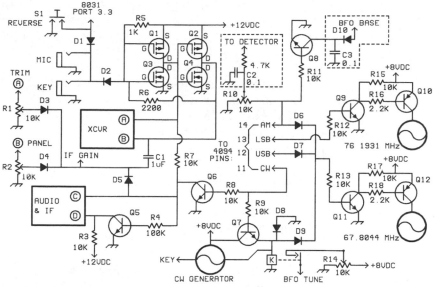

■ 5-24 *Schematic diagram of the control board. C1 is a nonelectrolytic type such as a plastic-film capacitor. All diodes are 1N4148 or similar silicon types. S1 is a momentary, normally open, push-button switch. Q1 and Q2 are IRF9510 vertical-channel metal-oxide semiconductor (VMOS) transistors. Q3 and Q4 are IRF510 VMOS transistors. All of the bipolar transistors can be any small, general-purpose silicon transistors. Just make sure they are of the correct polarity. Also make sure the leads are inserted correctly. Not all small transistor packages use the emitter-base-collector sequence; some use the base-collector-emitter sequence.*

If a partial ground occurs at the key from something such as dirty contacts, the signal applied to Q1 and Q2 could be somewhere between 12 V and ground. This can cause unpredictable operation. Resistor R6 helps the quad of MOSFETs attain a definite on or off state by providing positive feedback and a hysteresis effect. As long as the keying signal is high or low enough to force a sufficient change, positive feedback from R6 steps in to finish the job.

One of the events monitored by the microcontroller is the closure of a key or microphone switch. Diode D1 allows the placing of the synthesizer on the transmit frequency without actually transmitting by pressing S1. This means the operator can listen momentarily on the transmit frequency while in the transmit-incremental-tune mode (XIT) by pressing the reverse switch.

The audio and lower IF boards have T/R requirements that differ from most of the other circuits for some emission and reception modes. Specifically, these differences relate to the circled-C and circled-D control lines. Whenever the 4094 shift register outputs a high at pin 11 for CW operation, Q6 is turned on and forces the circled-D control line low during both transmit and receive operation. Transistor Q5 inverts whatever the Q6 collector is doing, whether in phone or CW mode.

Transistor Q7 is also energized for CW operation, supplying bias to the keyed CW crystal oscillator and powering a relay. Voltage from the relay contact is used to tune the BFO to a frequency suitable for CW reception. Current from Q7 is also used to turn on Q11 and Q12 to power the second local oscillator at 67.8044 MHz.

Mode commands from the 4094 shift register are sent by making pin 11, 12, 13, or 14 high and the other three low. AM and lower sideband also use the 67.8044 oscillator. Upper sideband turns on Q9 and Q10 for the 76.1931-MHz oscillator.

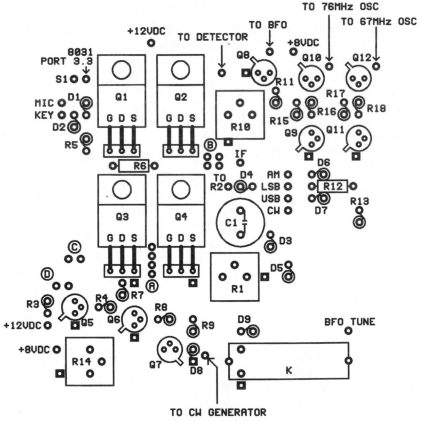

■ **5-25** *Layout for the controller board. The pads for input connections labeled AM, LSB, USB, and CW are spaced 0.1 in apart in case you want to use standard header pins and sockets here.*

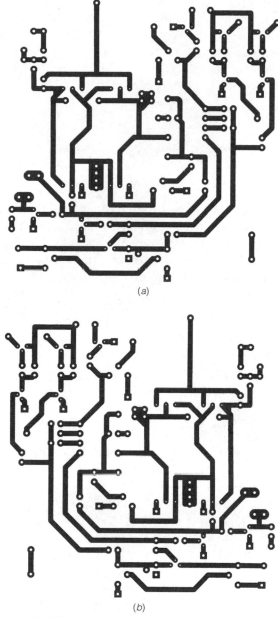

(a)

(b)

■ **5-26** *These are the etching patterns (actual size) for the control board. Use (a) for photographic reproduction, and (b) for the transparency iron method.*

Here I must make another confession. The AM-related components such as R10 and Q8 are suggestions for implementing AM receiving and transmitting capabilities. Unfortunately, I've not had the time to experiment with this mode yet. What little amateur and broadcast AM I've listened to was accomplished by zero-beating signals in an SSB mode.

The idea behind R10 is to unbalance the product detector/balanced modulator. This will allow it to detect and generate AM signals. The resistor and detector enclosed in broken lines (in figure 5-24) are to be located on the detector/modulator board. The resistor connects to the audio port of the detector.

Components in the other dashed-line enclosure in figure 5-24, consisting of D10 and C3, are to disable the BFO during AM operation. This is just one possible way of stopping the oscillations. A second NPN transistor and pull-up resistor driven by Q8 could be used to control the supply going to the BFO.

Figure 5-24 is a rather simple control scheme and works well. However, it does little to time the sequence of events as the operator changes from receive to transmit and back again. The main objection to changing everything at once is an audible pop or thump. Capacitor C1 helps alleviate the problem. Understanding this circuit is a good start for those who want to design a different control board. Quieter break-in operation should be possible with a slightly more complicated circuit.

Parts located on the control board are shown in figure 5-25. Transistors Q1 through Q4 do not need heat sinks for this circuit. Other transistors are shown with metal-can outlines, but plastic-case types are just fine. This circuit can be built on a conventional printed board, as shown in figure 5-26. It is also a good candidate for perforated board construction. A perforated board allows you to easily add to or change the circuit.

Signal generation

6

WITH MOST OF THE ANALOG CIRCUITS NECESSARY FOR processing radio signals completed, two building blocks remain in the completion of a radio transceiver capable of functioning over a large frequency span. One is a wide-range oscillator or signal source. The other is a controller, in this case a microcontroller. Covering wide expanses of the radio spectrum can be cumbersome for the operator if he or she must make several adjustments each time a substantial change in frequency is attempted. To some extent, the up-conversion, broadband design used here helps because fewer circuits must be switched in and out.

One major section, the microprocessor-based controller, automates some of the chores which would otherwise be required of the operator. Switching input/output filters and controlling the frequency synthesizer allows operators to change frequency settings with greater alacrity than many older designs constructed in home workshops. The microcontroller handles a few other chores, such as mode selection. As you will see in later chapters, most of its time is spent managing operating-frequency parameters.

The other major building block is the synthesizer, which must output a stable, predictable signal of reasonable accuracy and good spectral purity. Some parts of this synthesizer are made of analog circuits, contrasted with what would usually be thought of as digital circuits. Some fairly conventional oscillators are influenced by the output of digital counter circuits in the upcoming synthesizer.

Although there is great progress being made in direct digital synthesis (DDS), analog oscillators used in indirect synthesis will probably be around for some time because of the difficulties of implementing direct digital methods at higher frequencies.

Early signal generators

Before getting into the intricacies of fulfilling the requirements of a signal generator (synthesizer) for your transceiver, I thought it might be fun to look back at some early methods for generating signals used in radio communication. This has no direct application to your transceiver project; however, it does point out how technically inclined amateurs can now construct signal-generating apparatus with performance and features unattainable at any price in bygone years.

One of the first methods of generating high-frequency currents necessary for radio communication was electrical spark discharge (see figure 6-1). This was the method used well before vacuum tubes existed.

The basic spark-gap transmitter consisted of electrodes separated by a gas—usually air—with a high-voltage source applied when keyed. When a voltage of sufficient intensity is applied to air or other gases, electrons are stripped from molecules as the gas becomes ionized. With a conductive path formed through the gas, it becomes heated and expands. Drastic undulations in the current occur at a rapid rate. These somewhat chaotic variations were used to shock the antenna circuit into damped oscillations, thereby radiating a signal into space.

■ **6-1** *Manmade spark-gap transmitters are not totally extinct. You may experience the unfortunate necessity of tracking signals from accidental sources such as this electric fence charger.*

At first, antennas were connected directly to the spark gap. Later refinements included adding selective elements such as inductive coupling with tuned circuits, which concentrated more of the energy into a narrower portion of the spectrum. This was an improvement, but nothing near the narrow spectrum emitted by even early vacuum-tube CW oscillators. Close-by receivers suffered spark interference over hundreds of kilohertz.

Other methods of generating radio-frequency energy included mechanical devices such as alternators. You can imagine the rotational speeds necessary for generating radio-frequency current. Although this method is certainly capable of greater spectral purity, mechanical stresses limited it to the lowest frequency ranges.

Vacuum tube oscillators were a major advance in the ability to concentrate energy on a single frequency. Tubes were also the first significant devices to be used as signal-generating devices in receivers. The circuit known as a *regenerative detector* could be operated in a manner such that its own weak oscillations mixed with incoming CW signals to produce an audible beat note in the headphones.

Those who are familiar with these historical events realize I am glossing over and ignoring some very interesting people and events. It would be interesting to say more about what happened, but I'm probably stretching the scope of the book enough in this chapter. See sources listed in the Bibliography for more information.

Vacuum tubes

Also known as valves, early tube transmitters were basically an oscillator connected to an antenna. The antenna was usually a major part of the oscillator circuit, as shown in figure 6-2. This, I suppose, was a carryover from spark techniques. Frequency stability was not what we are familiar with today. With the tubes available for these early transmitters, it was an accomplishment to sustain oscillations and deliver power to the antenna.

Eventually, oscillators were used to generate a relatively low power signal which was amplified by other tube stages. Successful applications of this master-oscillator power-amplifier (MOPA) design finally became practical when better methods of building RF amplifiers were developed. Isolating the oscillator provided better thermal stability because the stage could run at lower power levels and with less heating of frequency-determining com-

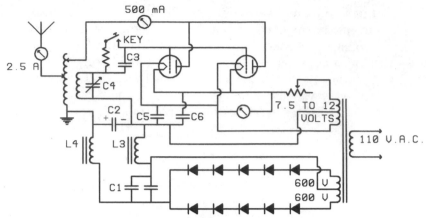

■ **6-2** *This two-tube oscillator, along with those shown in figures 6-3 and 6-4 are redrawn from illustrations in* Fifty Years of A.R.R.L. *1st ed. (1965). (Courtesy of American Radio Relay League)* Modern component symbols are used here. The rectifier bank is made of lead and aluminum electrodes immersed in jars filled with a borax or ammonium electrolyte.

ponents. Changes in loading at the antenna caused by swaying or other disturbances had less effect. With enough stages and sufficient isolation between the oscillator and antenna, an amplifier could be keyed instead of the oscillator and, hopefully, better keying would result.

The ultimate signal source for stability was the crystal-controlled oscillator. Figure 6-3 is an example of a popular circuit. The transmitter usually functioned either at the crystal frequency or a harmonic. Quartz-crystal oscillators are used as frequency-determining elements in most of the amateur radios manufactured today, but in a different way.

Crystal resonators are used as a stable reference against which the frequency of a variable-frequency oscillator is compared for indirect synthesis. They are also used as clocks in a digital-to-analog conversion process which builds waveforms of the desired frequency in a step-by-step fashion called *direct digital synthesis.*

Early tube oscillators have much in common with the solid-state oscillator circuits used in this book, even though the active devices are certainly different. Both types rely solely on properties present in the oscillator for maintaining a constant frequency. When more stringent requirements, such as a large tuning range, higher-frequency operation, and low drift, must be met, additional techniques must be used. Before going on to how this is done, let's take a quick look at how information was conveyed on early radio signals.

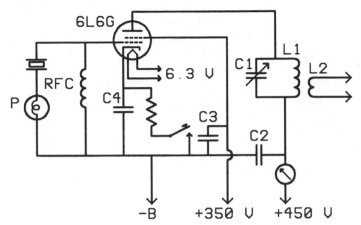

■ 6-3 *A one-tube crystal-controlled transmitter popular in the late 1930s and early 1940s known as the QSL 40. P is a small incandescent indicator light.* (Courtesy of American Radio Relay League)

Modulation

The simple presence of a signal does not by itself convey much information. In most cases, the signal must be changed or modulated in some way to be useful for communication purposes. The earliest regularly used method was on/off keying of the transmitted signal. Of course, this was an early form of digital communication with the data processing hardware located between the operator's ears instead of inside a desktop box.

On/off keying was used with spark, and later with tube-type, transmitters. Around 1906, voice transmission was even accomplished with spark. Soon, improvements were made in transmitting equipment, using an electric arc which produced a carrier signal quiet enough to allow reasonable sound fidelity (for the period). Unfortunately, these early attempts were very hard on some of the studio hardware. Here's why.

The carbon microphone is one of the earliest transducers for converting sound to electrical information. It is used to this day in many telephone mouthpieces. It is made of loosely packed carbon granules behind a flexible diaphragm. Electrodes are mounted in contact with the granules. Sound waves strike the diaphragm, causing it to exert varying pressure on the granules in step with the sound waves. As the granules are alternately squeezed and released, conductivity through the pack varies. An electrical waveform resembling the sound wave is produced.

This was, in essence, the modulation system. Megaphone-like cones were used to concentrate the sound. Even with acoustic reinforcement, announcers and musicians had to speak and play as loudly as possible. The unfortunate carbon microphone was forced to contend with a great deal of current. As one soon overheated and ceased functioning, another had to be substituted in its place. It must have been a lively experience for the sound engineer!

The ability of vacuum tubes to amplify greatly improved the situation. Microphones could be operated in a more reasonable manner. Tubes amplified the small audio signal to a level which could adequately modulate the RF carrier. An early amplitude-modulated transmitter might look like that in figure 6-4.

Direct synthesis

Between the early days of radio and the present, a lot of innovative signal-generating schemes have been developed. One of the most recent to see widespread use is direct digital synthesis. It is an extraordinarily flexible method. As stated earlier, it is a process of building a waveform bit by bit.

Instead of relying on the time constant of an LC or RC circuit to determine the frequency of oscillation, individual points in a waveform (sine, sawtooth, or whatever) are attained by setting a voltage at a selected point in the waveform for a small amount of time. During the next increment in time, another discrete point in the waveform is formed by again setting the voltage to an appropriate level.

This incremental method of synthesis produces a stepped approximation of the intended waveform, as shown in figure 6-5. This grainy-looking signal is then passed through a low-pass filter. The

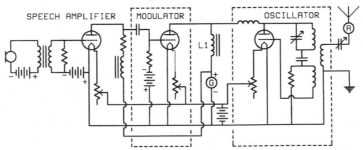

■ **6-4** *A 1921 circuit for amplitude-modulated telephony. In addition to the motor/generator plate supply, note the number of batteries used!* (Courtesy of American Radio Relay League)

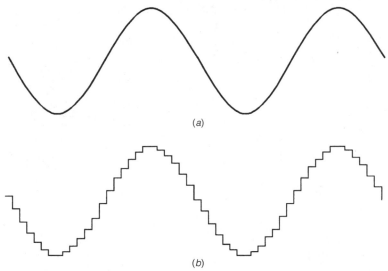

(a)

(b)

■ **6-5** *(a) The pattern of an ideal waveform we might wish to produce by direct digital synthesis. (b) An approximation produced by
the output of a digital-to-analog converter.*

cutoff frequency of the filter is low enough to pass the desired frequency but high enough to smooth all the jagged edges into a
waveform very close to the intended shape.

Individual steps in the waveform are produced by a digital-to-
analog converter (DAC) circuit. Normally, the digital information
going into the DAC is represented by simple binary signals in parallel form. Each line at the input represents a power of 2. If you're
not familiar with numbers represented in binary or base-2 form,
see chapter 12 for more information.

For now, let's assume the proper digital value for a particular point
in the waveform has already been generated. By some means, we
need to translate this value to a voltage. One simple way might be
to use the scheme shown in figure 6-6. Current from a known, stable voltage source is sent through any of the switches, which happen to be in the up position. If you would like to experiment with
a breadboard version of this circuit, you could use toggle switches
and supply the digital information manually.

Of course, you would be hard-pressed to flip switches fast enough
to synthesize anything other than extremely low frequencies. In a
practical DAC, each switch is formed by a circuit using multiple
transistors. Current flowing through the op-amp input is limited
by a resistor connected to each switch.

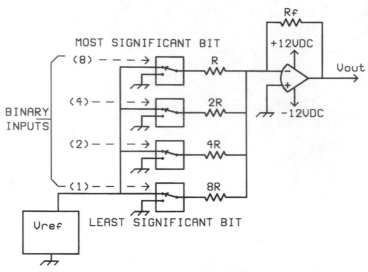

■ **6-6** *This is a straightforward way to build a 4-bit DAC. It is not easily expanded to handle larger numbers.*

Each resistor connected to a switch has a value which sets current flow corresponding to the binary bit controlling the switch. The op-amp circuit is a convenient way of adding individual currents contributed by each resistor and converting the total to a corresponding output voltage.

Remember that op-amps have extremely high gain. Voltage differences between the inverting and noninverting input will be practically zero because feedback through R_f will steer the output toward the voltage present at the noninverting input. In this case, it would steer the output to zero if no input from the switch resistors was present.

When current from one or more of the switch resistors is present at the inverting input, current will flow through R_f as the output voltage changes. Current from R_f, which should be ½R or less, will buck current from the switch resistors. It doesn't matter if V_{ref} is positive or negative, current from R_f will still hold the inverting input very close to the noninverting input. How far it gets forced away from that value depends on the open loop gain of the op-amp.

The end result is that voltage at the inverting input always looks like zero as the op-amp output swings its output voltage to counteract current from the switch resistors. You might say current from the switch resistors is fooled into thinking it's going to ground. Virtually no current flows into the noninverting input because of the very high impedance there. Practically all current to or from the switch resistors flows through R_f. The relationship between R_f and V_{out} is:

$$V_{out} = R_f * \text{current in } R_f \qquad (6\text{-}1)$$

Current in R_f is, for our purpose, the same as the total from the switch resistors.

A scheme like that shown in figure 6-6 might work for converting 4-bit numbers, but using larger numbers for greater precision is not practical. The resistors' values and tolerances become a major problem. A 4-bit converter as described needs values spanning an 8-to-1 range and tolerances of 1 part in 16 or better. This is within reason. A 16-bit converter would have unrealistic requirements. The resistors would have to be accurate to one part in 65,536 or better.

A more practical solution generates the binary weighted currents from a resistor network instead of relying on the accuracy of individual resistors. Figure 6-7 shows how this is done using only two resistor values. To understand how this works, notice that each $2R$ resistor is always connected to either ground or virtual ground at the inverting input terminal because the switches have only two

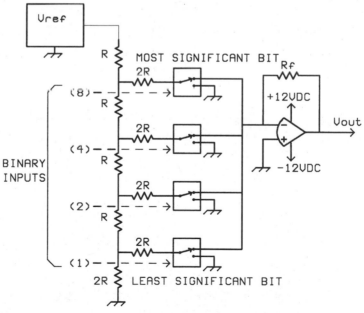

■ **6-7** *The* R-2R *ladder network is well suited for building larger DACs.*

possible positions. Regardless of which way the switch is thrown, the current path is effectively to ground.

Currents divide in half at each junction. The easiest way to see this is to look first at the least significant bit (LSB) corresponding to the one's place in binary numbers. When current flows into each $2R$ resistor, any current at the resistor junction divides equally because the resistances are equal. Now look at the next higher bit, the two's place. R, in series with the two parallel $2R$ resistors, forms a path with a resistance of $2R$. Same situation: the switch resistor and path below it divide the current equally.

Current is divided in the same manner right on up the R-$2R$ ladder network. To produce an accurate replica of a sine wave or other arbitrary waveform, the DAC must be able to emit voltages as close as possible to a level corresponding to an identical point on an ideal waveform at each designated point in time. DACs with more bits of resolution are needed for better waveform fidelity.

Building a more accurate replica of the intended waveform is also aided by setting the DAC to a new level more often. If one cycle of the waveform is formed by outputting 40 different levels, the waveform will obviously be smoother than one built from only four voltage steps. Fast, high-precision DACs can produce better waveforms, but they are also harder to produce and more expensive. As frequency, speed, and precision increase, other parts of a direct digital synthesizer also become more difficult and expensive.

Figure 6-8 is a block diagram of how the DAC and low-pass filter can be teamed with digital circuits to implement one kind of direct digital synthesis. I use a wide arrow pointing to the DAC to indicate multiple parallel lines for moving digital data. Some DACs have additional circuitry which can accept information in a serial form, one bit after another on one wire. In this application, it is more appropriate to use a fast parallel interface, with each bit assigned a single wire. When it's time to update the DAC, all bits are sent simultaneously.

Because the DAC output must change precisely on time, a signal from the clock is used. The clock is usually a crystal oscillator or other stable source. The clock signal may come directly from the crystal oscillator or may be an integral fraction of the oscillator frequency. The read-only memory (ROM), phase accumulator, and adder are all digital circuits. If you are not familiar with these kinds of digital circuits, just think of them for now as "blackboxes" which manipulate binary numbers.

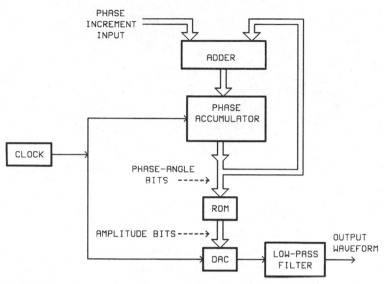

■ 6-8 *This is a block diagram of one possible scheme for generating signals by direct digital synthesis. The phase increment input is often generated by a microprocessor.*

Changing the DAC output on cue can be accomplished by using additional circuits to latch the information presented to its inputs. This additional circuitry is often built into the DAC. When a logic transition from the clock occurs, the latch changes output levels to match whatever binary values are present at its inputs. The latch then holds this state, regardless of what is sent to its inputs, until another clock transition occurs. This scheme helps eliminate glitches on the DAC output caused by information from ROM arriving at the wrong time.

Information is stored in ROM as numbers representing amplitude at many points on a sample waveform. Discrete points in time correspond to addresses in ROM. Each ROM address is described by a number. The ROM look-up table has amplitude information necessary to describe a waveform such as a sine wave. Although they don't have to be, ROM addresses can be accessed in sequential order. If we were to send all of the table entries in ROM to the DAC in a sequence timed by clock transitions, a waveform is produced. It would be the lowest frequency which the synthesizer is capable of producing for a given clock frequency.

The location of any point on the waveform is also described by the distance from the beginning/ending point of the waveform. The location along the horizontal or time axis is known as the *phase angle.* ROM addresses could correspond to phase angles ranging from 0° to 360°. If 360 table entries existed, each ROM address used in the table would represent a 1° phase angle.

A different number of table entries is often used, such as a power of 2. Each address would represent a 0.703125° increment in a 512-entry table. It is also possible to use a look-up table covering only the first quadrant (0° to 90°) and change signs to obtain values for other quadrants.

Higher frequencies are produced by selecting fewer ROM addresses. This is what the adder and phase accumulator are for. If every other address was used, the output frequency would be double the first example. Instead of jumping two addresses on each clock transition, you could set the phase increment input to jump three, four, or five increments, and so on. The phase accumulator remembers the last horizontal-axis location or phase angle on the waveform. The adder simply adds this number to the next phase increment to find the next ROM address.

Oops, no DDS

That's right. Now you may be wondering why I chose to drag you through all of this information about direct digital synthesis if it's not used in the transceiver. I have several reasons. The synthesizer scheme actually used here contains a DAC which is applied in a different manner. DDS hardware is becoming more available and some advanced experimenters may want to modify the transceiver design to use it. Like all forms of signal generation, DDS generates spurious components in the output spectrum, which you need to be aware of.

Output frequency for a DDS system like that in figure 6-8 is calculated in the following way.

$$\text{Output frequency} = \frac{\text{clock frequency}}{\text{total table entries} * N} \qquad (6\text{-}2)$$

N is the number of entries between one clock transition and the next. The system steps through the table, outputting an amplitude step at every Nth entry.

This system is limited in how well the waveform can be approximated because of limitations in DAC resolution and the number of amplitude steps which can be included in the output waveform. When N does not divide evenly into the table entries in a 360° time interval, another problem arises. Waveform distortion occurs because points at which amplitude steps occur on one waveform are different from the next. This produces some low-level sidebands with complex relationships to the programmed frequency.

Reasonably priced prepackaged systems that are suitable for amateur experimentation have been produced by some manufacturers. DDS is not a panacea. One product I'm familiar with has sideband levels low enough for a really nifty way to control a QRP (low-power) transmitter. However, I wouldn't want to use it in a high-dynamic-range receiver. The close-in spurs would, in some cases, cause false responses to strong signals.

DDS hardware should become more capable and affordable in the future. Producing signals by DDS at VHF and especially UHF upward is so difficult and expensive that other techniques will probably be used for a long time. I don't mean to discourage anyone from DDS experimentation. I'm simply advocating that you learn more about it when doing so.

A somewhat related acronym seen in amateur radio today is DSP, which stands for *digital signal processing*. Processing signals usually involves first converting analog signals to digital values (the reverse of what a DAC does), subjecting the digital information to various kinds of computations, and converting the digital information back to an analog signal.

DSP is often used to implement very sophisticated audio and even IF filters. It can even be used to make almost an entire transceiver, although that's not really practical at present. No, there's no DSP in the transceiver project. The good news is that installing audio DSP in your own version of the transceiver will be relatively simple. Commercial products, kits, and homebrew projects are available.

Most of the recently advertised DSP systems are self-contained. They could easily be included in the transceiver enclosure as long as you provide sufficient room. Good shielding and decoupling techniques will be necessary. Of course, I'm referring to developed systems. Designing your own firmware and hardware would be much more challenging.

Oscillator observations

7

DIAGRAMS OF NUMEROUS OSCILLATORS HAVE BEEN PRE-
sented. To finish the multiband transceiver, a few more are
needed. Requirements of synthesizer oscillators are in some ways
the most demanding of any oscillator in the transceiver project.
For this reason, I wish to examine some basic ideas and offer a few
personal observations.

For oscillations in a system to occur, we need a system for storing
and releasing energy at regular intervals. This is true whether
electrical, mechanical, or other forms of energy are involved. As
energy is stored and released, losses occur. Energy taken out of
the system to perform useful work will also reduce the strength or
amplitude of the oscillations.

Unless we are content with oscillations which dissipate over time
and become useless, energy must be put back into the system. A
swinging mechanical pendulum needs added acceleration to
regain energy lost as heat to materials and air friction. Just as well-
timed thrust keeps the pendulum swinging, electrical impulses
can sustain oscillations in various kinds of resonators.

Care and feeding of oscillators

A resonator such as a parallel-tuned circuit can support oscillations but only for brief periods. They quickly die out because of
losses in the circuit. Quartz crystal resonators exhibit much lower
losses. You can even hear the ringing effect in some radio receivers
as the vibrations fade away in a narrow crystal filter.

The usual method for sustaining oscillations is to use an amplifier to
sample energy from the resonator and feed back an amplified portion to the resonator in phase with the original signal. The resonator
is also a filter and is so labeled in figure 7-1. The filter in some oscillators can actually be a network with high losses, incapable of storing much energy as with some RC phase-shift networks. RF
oscillators usually need as little loss in the filter as possible.

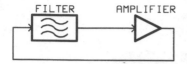

FILTER   AMPLIFIER

■ **7-1** *Part of the output signal must arrive in phase at the input for a noninverting amplifier. The filter may include impedance-matching properties.*

All sorts of devices can be used for the amplifier. For most RF equipment, including the transceiver, the amplifying device is a single transistor. Exceptions do exist. Integrated and discrete versions of multitransistor oscillators have been used, usually when some special requirement exists in addition to emitting a good, clean signal.

In order for oscillations to continue, amplifier gain must exceed losses in the circuit. Unless the total gain and loss are equal, amplitude of the oscillations will continue to grow forever. Obviously, that doesn't happen. After the amplitude reaches a certain level, gain begins to decrease. Further amplitude increases cause gain to drop even more. At some point, amplitude is limited because gain and loss in the circuit are exactly equal.

I've chosen to use bipolar transistors; however, field effect transistors are also used in circuits popular with amateur experimenters. The usual method of amplitude limiting is different for the two types. With FET oscillators, a reverse-bias voltage is allowed to build on the gate as amplitude increases. As the FET is pushed closer to pinch-off, gain decreases.

Bipolar transistor oscillators limit because of nonlinearity. Bipolar stages are normally biased for constant emitter current. As RF drive to the base increases, the output signal is increasingly distorted. A larger percentage of the output power is contained in harmonics, while less is present at the fundamental frequency.

Practical examples of circuits like that shown in figure 7-2 were illustrated in earlier chapters. It provides reasonably good performance. It also seems to be one of the easier circuits for neophytes to successfully build. This may seem like a strange statement. If you build an oscillator exactly as instructed in a published design, would it not work as well as the original? It may, or it may simply sit there looking like a useless collection of parts while you ponder what went wrong.

Failure to oscillate can happen for a number of reasons. I've heard builders complain, "All component values and connections are the same. Why doesn't it work?" The obvious reason is that something is different. Unloaded Q of the original tuned circuit is often not specified. Transistor gain in the prototype may not be given.

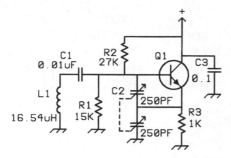

7-2 *This grounded-collector Colpitts oscillator is designed for a lower frequency limit of approximately 3.5 MHz, as are figures 7-4 through 7-7. Supply could be at or around 8 V and should be well regulated for all the examples.*

Authors of various circuits didn't design them simply to demonstrate their skill at thwarting repeatability. Usually the designer is trying to enhance frequency stability in some manner. This may include long-term stability effects as in drift or short-term stability as in noise. Circuits such as those in figures 7-3 and 7-4 have attributes which have been used to advantage for both purposes.

One advantage to be gained is a better impedance match presented to the transistor. Another is higher resonator Q. If not for

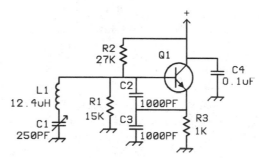

7-3 *A series tuned Colpitts, also known as the Clapp configuration.*

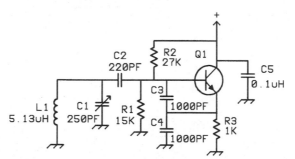

7-4 *A grounded-collector version of the Seiler oscillator. C2 in combination with the other reactive components is small enough to provide an impedance transformation.*

series capacitor C2 in figure 7-4, feedback capacitors would have to be smaller. Capacitance changes at the transistor terminals caused by load or supply variations affect the frequency of oscillation less if the C2 and C3 of figure 7-3 are relatively large. Capacitors C3 and C4 have the same effect in figure 7-4. This arrangement has also been used with FETs, as in figure 7-5.

Capacitors for coupling and matching, such as C2, can be only so small. A designer who trudges off too far in this direction may reach a point where sample-to-sample variations in components cause problems when trying to duplicate the circuit. Field effect transistors are especially prone to variations in transconductance from one unit to the next. The usual small FET used in oscillators does not have as much power gain at RF as some bipolar transistors. I've talked to experimenters who have built nonfunctioning FET Colpitts and Clapp oscillators from published designs.

The circuits finally work when a "hot" enough transistor and/or a low-loss inductor is substituted. Figure 7-6 is one FET circuit which practically every builder has good luck with: the Hartley. Figure 7-7 shows that bipolar transistors can be used in this circuit as well. The problem with Hartley oscillators using bipolar transistors is—well, you might say they are overly eager.

Feedback through the tapped inductor L1 can occur at frequencies far below the operating frequency, where transistor gain is much higher. When this happens, the oscillation may occur at tens or hundreds of kilohertz and at the resonator frequency simultaneously. This is likely to happen if C2 is too large. This behavior is often referred to as *squegging*.

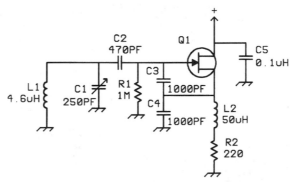

■ **7-5** *Another Seiler circuit with an FET. R2 is not necessary in many cases because of the negative dc component, which builds up at the gate and acts as an automatic gain control mechanism.*

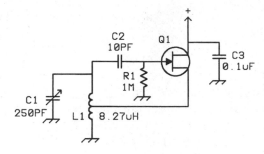

7-6 *A Hartley oscillator. The tap on L1 is often at one-quarter to one-third of the winding from the grounded end.*

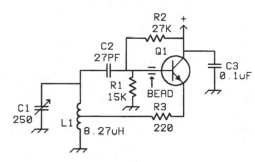

7-7 *A bipolar version of the Hartley.*

A smaller C2 in the Hartley circuit also reduces loading on the resonator and enhances Q. Some published circuits seem to have been refined to the point that the amount of feedback is marginal.

One method of reducing the problem is to use a smaller C2. You can also increase the value of R3. Parasitic oscillations can also occur at VHF and UHF in any oscillator using a transistor with a high gain-bandwidth product. Placing a high-permeability ferrite bead on the collector or base usually eliminates the problem

An output signal can be extracted from these and other oscillator circuits in many ways. Usually a small-value coupling capacitor is connected to the emitter or source to drive a buffer amplifier. Regardless of the method used, the external circuit should affect the oscillator as little as possible.

Power taken from the oscillator should be as small as possible but not so low that amplifier noise overrides oscillator noise. The load presented by the amplifier input should not vary.

Tuning

Most LC-tuned oscillators built by amateurs are designed to change frequency by varying capacitance in the resonator. High-quality air-dielectric variable capacitors are no longer easily procured by many experimenters. Technological change has, in effect, eliminated much of the demand for such components.

Polyethylene-film variable capacitors are perhaps the most common of currently available variable capacitors designed to be adjusted from the front panel. They don't have the stability or unloaded Q of air variables. Not all low-end and midrange broadcast receivers are synthesized yet, so maybe the diminutive variables will be around for some time.

Of course, you can also change frequency by varying inductance. Some commercial producers of amateur gear have used movable powdered iron cores inside specially wound inductors for frequency control. The problem with variable inductances is that none are readily available to amateur builders. At least not any that I know of which are designed to be manipulated from the front panel. Building your own is possible, although achieving good electrical and mechanical stability would be very challenging.

The task of tuning a radio-frequency oscillator is now often handled by varactor diodes. The PN junctions of these solid-state devices are used as a variable capacitor. Actual capacitance value depends on the reverse-bias voltage present across the junction. Lower voltages result in greater capacitance; higher voltages do the opposite. Unloaded Q varies with the reverse-bias voltage, rising with higher voltages. Unloaded Q decreases as the operating frequency is increased.

Ruggedness along with small size and cost make varactor diodes very appealing in some cases. They are well suited to some tasks which don't require high stability or can be stabilized with a feedback loop. Unfortunately, varactors have much lower thermal stability than a good air-variable capacitor. These diodes often have much lower Q than conventional capacitors and contribute to the noise content of oscillator signals.

Some amateurs who wish to build high-quality tunable oscillators might interpret the previous information as cause for pessimism. In fact, you can achieve good to excellent results in the home workshop using rather mundane components. To do this requires making the correct compromises and using feedback for stabilization.

Noise

Unfortunately, all oscillators have noise in their output signals. Noise traveling through the filter or resonator comes out with its energy concentrated into a narrower spectral band than when it went in. As a signal is fed around the loop, it becomes more tightly concentrated about a single frequency. The iterative nature of this action actually filters out noise to a much greater degree than the filter could by itself. Nevertheless, an output signal can be viewed as a well-filtered noise source, as illustrated in figure 7-8.

The noise content of most RF oscillators was low enough to avoid much concern before the 1980s—at least in the amateur community. Two or more trends conspired to focus more attention on it. Manufacturers began using frequency synthesis, and receiver circuits capable of wider dynamic range were becoming more common. A third factor is that you could probably find more stations using high power and large antenna arrays than in earlier eras.

When mixer and amplifier circuits were developed to cope with a larger range of signal strength, oscillator noise sometimes limited the ability to handle strong nearby signals. A receiver which could otherwise function well, tuned close to strong signals, may receive apparent interference because of noise from its own oscillators.

Here is why noisy oscillators cause problems when receiving. The center of the oscillator signal mixed with a strong nearby signal does not produce a mixing product which falls within the IF passband in this example. Such things can happen but it is a result of intermodulation distortion, an event originating usually in mixer or amplifier circuits. In instances when the receiver is limited by oscillator noise, noise sidebands, offset from the oscillator center-frequency, mix with the strong signal. The operator is then plagued with interference that would otherwise be blocked out by selective IF filters.

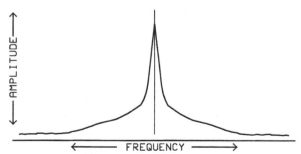

■ 7-8 *The centered vertical line represents an ideal, although impossible, signal with all of its energy concentrated at one frequency. Real oscillators have their output energy spread over a range of frequencies.*

For this reason, a noisy oscillator can ruin the capability of an otherwise excellent receiver to ignore nearby strong signals. No matter how well the remainder of the receiver circuits function, excessively broad noise sidebands from the first local oscillator broaden the response. The noise problem is even more serious if it occurs in a transmitter. Other stations tuned in near the noisy signal will then receive interference.

Many people refer to the previously described problems as *phase noise*. This is because random variations in phase are usually the major contributing factor in the symptoms described above. Many commonly used mixers are relatively insensitive to amplitude variations in the oscillator signal because they are operated in a saturated switching mode.

Amplitude variations in sinusoidal signals can be converted to variations in phase. Sine waves are converted to square waves by the switching action of the mixer or, in some cases, amplifier stage. Switching points on the sinusoid are affected by variations in amplitude. The net result is that random amplitude variations of oscillator output are converted to random phase variations.

I mentioned frequency synthesis as one of the factors which contributed to increased awareness of oscillator noise problems. To be blunt, some of the earlier synthesizers had relatively noisy outputs. Along with random noise, some early systems had other spurious output components, which I will mention later. Highly integrated components currently available make it easier to build indirect synthesis systems.

Varactor tuning diodes are a wonderfully convenient way to remotely tune an oscillator. A single conductor can bring in the reverse-bias voltage needed for the tuning diode. No mechanical linkage is needed for tuning.

Perhaps the mechanical superiority of tuning diodes is sometimes too alluring. Many tuning diodes have enough capacitance range to allow a voltage-controlled oscillator (VCO) to cover a frequency span of 2 to 1 or more. High-frequency radio applications are usually not a good choice for such oscillators.

The safest bet for amateur builders is to restrict the diode tuning range as much as possible. If more range is needed, other measures should be used. You can switch capacitance and/or inductance values, as shown in figure 7-9. Multiple oscillator circuits are another possibility.

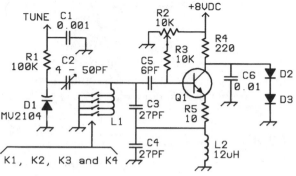

■ 7-9 *This VCO circuit is one which was actually used in the transceiver project at one time. Q1 is a BFR90. The frequency of operation ranged from 72 to 103 MHz. K1 through K4 are reed relays controlled by a microcontroller interface.*

If C2 is large, most of the RF voltage developed across the resonator will be applied across tuning diode D1. Reverse bias for D1 may not be high enough to prevent it from being driven into conduction during some parts of the RF cycle. When this happens, noise performance is severely degraded.

You could run the oscillator at a reduced supply voltage. Lowering power and voltage in the resonator may stop conduction, but it is a poor compromise. Decreasing resonant circuit energy will also reduce the signal-to-noise ratio of the output signal.

Restricting the tuning range of a VCO has another advantage not related to noise generated within the oscillator. A restricted range means less sensitivity on the tuning line. Larger variations in control voltage are needed for a given frequency change.

This is important because VCOs are often used in conjunction with digital circuits, which generate various types of switching impulses. A VCO capable of tuning 20 MHz with a 5-V change on the tuning line has a sensitivity of 4 MHz/V. Only a 2.5-µV change is needed to make the signal deviate 10 Hz.

Even a 250-µV signal imposed on the tuning line would be hard to see with most oscilloscopes. That is large enough to cause major problems because of the frequency modulation it would produce.

Minute signals can be coupled into the tuning line by stray capacitive or inductive coupling. Ground loops, in which currents returning to ground are unbalanced, are a potential problem. Ground bounce is also troublesome when fast-rising signals develop voltage across an imperfect ground.

Use a tuning diode for as little tuning as possible if you want to reduce the trials and tribulations of building a clean VCO. Make C2 as small as possible. If any additional tuning is needed, use other means to change inductance and capacitance values.

Construction and frequency stability

All RF oscillators, especially LC-tuned types, should use solid mechanical construction. Flexible wire leads, which can move slightly when vibrated, may cause unwanted frequency modulation. This may be true even if they are not in a frequency-determining portion of the circuit. Any conductor close enough to couple to the resonant circuit may affect the frequency if it moves.

Accidental interaction may occur between an oscillator and other circuits if it is not housed in a shielded enclosure. Fields from powerful transmitter circuits can affect the operating frequency. When receiving, an unshielded oscillator may radiate signals into other circuits, causing spurious responses or desensing.

Not everyone wants to build a full-blown synthesizer for his or her next radio project. Air-dielectric variable capacitors are still carried by some small suppliers. Supplies may be uncertain and the selection limited, as one outlet ceases operation and another begins. Determined experimenters usually find a source. Flea markets and hamfests are another place to find usable variable capacitors.

So what are the options in building a conventional variable-frequency oscillator with no corrective feedback? The good news is that a careful builder probably won't have much trouble with noise problems. Most people who can build equipment good enough to be limited by oscillator noise will also learn how to build suitable oscillators. The bad news is that you may have to deal with thermally induced drift. Sources listed in the Bibliography are helpful in both cases.

In the next chapter, we take a look at the phase-locked loop method of frequency synthesis. The trend in technology today is to use feedback referenced to a stable source for frequency control. The next chapter also contains some ideas for correcting oscillator drift that lie somewhere between programmable synthesizers and free-running oscillators.

Indirect synthesis

A SIMPLE EXPLANATION OF THE PHASE-LOCKED LOOP (PLL) is that it's a means of controlling the frequency of a VCO by comparing it to a reference signal. If the comparison shows a discrepancy in phase or frequency between the two signals, a resulting error signal is used to tune the VCO in a direction needed to reduce error.

A stable source such as a crystal oscillator is often used to generate the reference frequency. Figure 8-1 shows how the VCO is effectively locked to the frequency of a crystal oscillator. Any thermally induced frequency drift in the VCO is canceled by feedback from the phase detector. Although this scheme limits the VCO to a single frequency, additional techniques allow us to use other numerical relationships between reference and VCO frequencies.

Feedback and the phase-locked loop

Before we find out how to tune a PLL to multiple frequencies, let's see how the phase error could be generated. Diode mixers can be used to detect phase differences between two different signals. The classic double-balanced mixer circuit shown in figure 8-2 has a dc component in the signal present at the IF port.

When signals at the local oscillator (LO) and RF port are 90° out of phase with each other, the dc component from a perfect double-balanced mixer will be zero. Real mixers may have a small offset voltage. Maximum output occurs at 0° and 180° phase difference between the two signals. If the maximum positive voltage is at 0°, the maximum negative voltage will be at 180°. Reversing the phase at RF or LO will reverse this relationship.

To be useful in a PLL, the IF-port signal must be filtered. The filter is a low-pass type and allows the error signal to tune the VCO. If the VCO and reference frequencies are the same with only a phase difference, the loop easily works to reduce the error signal.

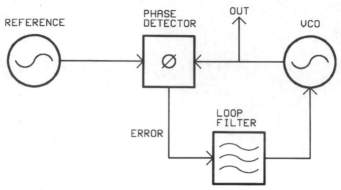

■ **8-1** *An example of how the VCO frequency and reference frequency could be locked together. One practical application for a system like this is as a cleanup loop. Spurious emissions from a dirty reference source can be eliminated by the loop filter if they are far enough away from the reference fundamental.*

A locked condition is achieved. If the frequencies are slightly different, like any good mixer, it generates sum and difference frequencies.

The sum is no problem because it is greatly attenuated by the low-pass filter. The difference frequency component, if it is low enough, rides through the filter with little attenuation. Modulated by the error signal, the VCO is swept up and down in frequency. If the VCO frequency comes close enough to the reference (or its harmonic), the loop will capture the VCO and regain lock. If not, the loop may remain unlocked until some means is used to bring the VCO within capture range of the reference.

Diode-mixer phase detectors are just one of several kinds of phase detectors. The diode mixer can also be used as a phase detector when one signal is an integer multiple of the other one. In this way

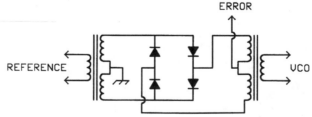

■ **8-2** *In addition to being a good mixer circuit, the diode-double-balanced mixer is useful as a phase detector capable of great bandwidth.*

the VCO can operate at a multiple of the reference frequency while locked to the reference harmonic by the loop. As an example, you could use a 100-kHz reference to lock the VCO to some multiple.

You can use the ability to lock on frequencies which are integer multiples of the reference to build a simple PLL system. A manually tuned variable capacitor can be used for coarse-tuning the VCO from one harmonic of the reference frequency to the next. The loop can be broken to tune between harmonics or, if the loop gain and capture range are small enough, phase lock will be broken as the VCO frequency is forced away by the manually tuned capacitor. When the VCO is close enough to another harmonic of the reference frequency, the loop will lock in.

Figure 8-3 is a photo of a multiband double-conversion super-heterodyne receiver which uses this type of synthesizer. The VCO can be locked at 100-kHz intervals. A tunable second IF covers 8.2 to 8.3 MHz. Instead of using numerous crystals, the VCO is coarse-tuned and locked to the appropriate frequency.

Some of the phase-detector output is sent to the input of the receiver's audio amplifier. When the coarse-tune control is used to

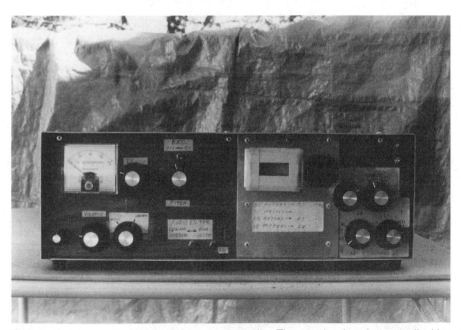

■ **8-3** *A strange-looking receiver, but it works. The synthesizer is controlled by the two upper-right-hand knobs. The far-right knob moves a variable capacitor for selecting the desired reference harmonic. The knob adjacent to it selects assorted VCO frequency ranges via a rotary switch.*

move the VCO near a reference harmonic, the operator hears the difference or beat frequency. As the VCO is manually tuned closer, the beat note suddenly disappears and the system is in lock.

Although this system beats purchasing dozens of quartz crystals, it certainly has its limitations. The loop cannot correct for VCO drift of more than several kilohertz or so. It sometimes loses the locked condition at higher frequencies. A 100-kHz component is still present in the error signal even after filtering. This frequency-modulates the VCO and causes spurious sidebands. Although all PLLs have some reference sidebands, other systems suppress them better than this one.

A logical way to start

Other kinds of circuits can detect not only phase errors but frequency errors as well. Digital electronic circuits are usually used for this purpose. Instead of processing signals as a continuous range of values, digital circuitry has a black-and-white view of the world. To circuits which process binary values, signals are either on or off, high or low. At least in theory, no gray areas exist. A value is either true or false. Rules of logic are then applied to perform useful operations on these dual-valued signals.

It is not hard to see how a continuous-valued signal such as a sine wave can be converted to high and low levels. If values above and below some point on a sinusoid are amplified and limited, the signal begins to take on the shape of a square wave. Using positive feedback in the limiting amplifier can help square the waveshape even more.

Various types of logical decision-making circuits can be used to count these square-wave pulses. A small number of logic elements are used as the basic building blocks of circuits which control and process digital data. If you like working with discrete transistor circuits in radio projects, you may find it interesting to look at logic elements and gates at the transistor level. A control circuit board mentioned earlier in the book uses circuits similar to those examined here.

I don't want to mislead anyone. This is simply a brief introduction. It may help you to understanding some of the concepts in this chapter, but I highly recommend using other resources to attain a more useful level of knowledge.

As mentioned, binary digital circuits deal with only two values. All sorts of voltage levels have been used to represent logic levels. For our discussion, let's assign anything between 0 and 2 V a value of 0. A signal between 3 and 5 V could represent a binary value of 1. Anything in between 2 and 3 is undefined—a situation you want to avoid in digital circuits.

Suppose the logic circuits use a 5-V supply for V_{cc}. A system in which the transistors are either saturated or turned off can have high and low logic levels very close to 0 and 5 V. The sample circuits to be presented can come within a fraction of a volt in meeting these limits, depending on component values, loads, and other factors.

One of the most basic operations you can perform on a logic level is to invert it. Assuming transistor gain and resistor values are appropriate, the following happens: if a high is present at the input shown in figure 8-4, collector current increases and the output drops to a few tenths of a volt. A low at the input lets the output voltage rise because practically no collector current is flowing. A circuit like this is sometimes referred to as a NOT gate.

OK, so no big deal, this is the same sort of thing that happens in a common-emitter analog amplifier. The phase is inverted. Maybe figure 8-5 will seem more impressive. It's a circuit which can actually make a decision. It's called an OR gate. It does just what the name suggests. If input A or input B or input C is high, the output is high. The output is low only when all inputs are low.

If you use an OR gate to drive an inverter, as in figure 8-6a, the logic function is that of a NOR gate. In this case, the output is high only if all inputs are low. One of many other ways to implement a NOR gate is shown in figure 8-6b.

Another very useful function is performed by the AND gate. In figure 8-7, input A and input B must be high to obtain a high output. Any other combination of inputs will result in the output being low. This gate, as well as the others, can be expanded to use more input terminals. You could have 3, 4, 5, or more inputs.

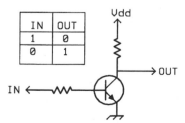

IN	OUT
1	0
0	1

■ **8-4** *An inverter or NOT gate. The list of input and output values is called a* truth table. *A high or low logic level is sometimes referred to as being true or false.*

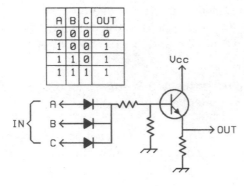

A	B	C	OUT
0	0	0	0
1	0	0	1
1	1	0	1
1	1	1	1

■ **8-5** *This is a three-input OR gate. The OR function is performed by diodes. The transistor simply provides higher output current than might otherwise be possible.*

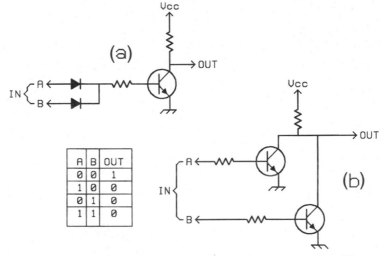

A	B	OUT
0	0	1
1	0	0
0	1	0
1	1	0

■ **8-6** *Both circuits are two-input NOR gates. It's just two different ways of accomplishing the same result.*

Building on the circuits already shown, a NAND gate can be implemented by using inverters to drive the inputs of an OR gate (see figure 8-8). The only input combination allowing a low output is when all inputs are high. Any other input combination will result in a high output.

Don't expect to see a lot of equipment built from circuits like those we just examined. Anyone who now manufactures logic circuits will use techniques which give much higher performance. It is possible to achieve higher speed with lower power dissipation if other schemes are used. Let's not get too bogged down in the details of how to do this.

It's more important to grasp the basic ideas than to have your interest obliterated by a massive infusion of information that the

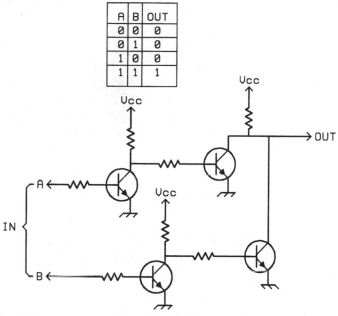

A	B	OUT
0	0	0
0	1	0
1	0	0
1	1	1

■ **8-7** *A two-input AND gate. AND gate truth tables are easy to remember. For inputs A through N, inputs A and B and C . . . and N must all be true to obtain a high or true result.*

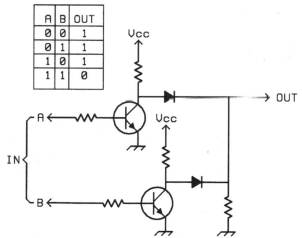

A	B	OUT
0	0	1
0	1	1
1	0	1
1	1	0

■ **8-8** *A two-input NAND gate. Same truth table as the AND gate except all inputs must be high to obtain a low output.*

IC designer needs. The preceding examples are used to illustrate a point because they are relatively easy to understand.

Circuits similar to these are examples of early solid-state logic known as *diode transistor logic* (DTL) and *resistor transistor logic* (RTL). DTL and RTL are now relics. Because of other logic families now available, such techniques are seldom used except in special cases like that of the controller board shown in figure 5-24. In that instance, only a few functions were needed. It was easier to deal with the various voltage and current requirements using discrete transistors than to use standard integrated circuits hooked to assorted interfaces.

Transistor transistor logic (TTL) and *complementary metal-oxide semiconductor* (CMOS) *logic* make up two of the more popular IC logic families. Many other families exist, including *programmable logic*. Transistors on a TTL IC chip are formed and connected in clever ways, which allow transistors to take over some of the work performed by diodes and resistors. CMOS is, for the most part, made of insulated-gate field-effect transistors.

Let's see how basic building blocks or gates can help build more capability into the phase-locked loop. At first glance, logic gates may not seem to be of much use in making decisions concerning a single signal source. After all, the decision-making ability of the previous examples relied on the combination of two or more inputs. These gates are known as *combinatorial logic elements*.

How to remember pulses

Combinatorial logic can act only on present input conditions. Sequential logic can remember a previous input state when it processes information in the present. You can build sequential logic elements from gates. The ability to act on past events can be used to count VCO cycles.

Flip-flop circuits are the most common type of basic sequential circuit. Flip-flops are not normally wired together from gates, at least in nonprogrammable ICs. You can buy individual flip-flops and more complicated circuits made from them in integrated-circuit form. However, we can use logic gates to build an understanding of how flip-flops work.

Figure 8-9 shows the logic in a reset/set (RS) flip-flop. With a high applied to the set input and a low applied to the reset input, the Q output will be high. The not-Q output will be low. If the set input

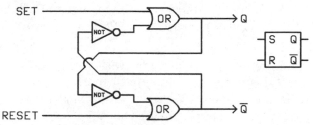

8-9 *This rendition of RS-flip-flop logic is done with American standard symbols. They don't usually have names printed on them. The shape is supposed to indicate their function.*

was to be made low again, the output terminals would remain unchanged. They would, in effect, remember the previous condition when the set input was high. You can verify this by tracing through the drawing.

This is a step closer to acting on a series of pulses, as mentioned earlier, but the RS flip-flop has two inputs. Figure 8-10 shows how some added logic will allow a single input signal to control the Q and not-Q outputs. This is one type of toggle flip-flop. Not only does it change outputs, but the change is dependent on the previous output state.

If the RS flip-flop in figure 8-10 is set, the AND gate feeding the set input will have one low input. Regardless of what happens at the toggle input, this top AND gate will output a low because both inputs must be high for the output to be high. The bottom AND gate, already supplied with one high input, lies ready. When the toggle input becomes high, the bottom AND gate outputs a high level and resets the RS flip-flop.

Nothing new happens at the output terminals when the toggle input goes low. The next time it goes high, conditions are different

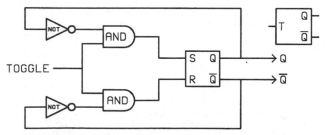

8-10 *This is just one of several ways to make a toggle flip-flop.*

at the AND-gate inputs. This time, only the top AND gate can produce a high at its output. The RS flip-flop is again set.

A significant fact about this sequence of events is that the toggle flip-flop reverses its outputs with every high-going input pulse. A high logic level at the output occurs only half as frequently as a high at the input. If a square-wave signal is applied to the toggle input, the output terminals will emit a similar signal at exactly half the input frequency.

You can further divide the input frequency by cascading flip-flops. Toggle flip-flops in figure 8-11 are drawn with small circles at the inputs. This means the same thing as an inverter or NOT gate which has been incorporated into the flip-flop. The inverters aren't needed for frequency division. They cause the output sequence in the truth table of figure 8-11.

Timing diagrams are illustrated for the input along with outputs A, B, and C. An external reset line is added to allow all flip-flops to start out in a reset state. A small arrow on the input waveform indicates the first positive-to-negative transition. It triggers the leftmost flip-flop into a set condition.

When two such transitions at the input have occurred, the leftmost flip-flop goes from positive to negative for the first time. This is indicated by another arrow on the A waveform. The same process is repeated at the B and C outputs.

Transitions are not instantaneous as the timing diagram might imply. The timing diagrams indicate what an oscilloscope trace might look like if the circuit is operating far below its maximum speed. A slight delay exists between input and output actions.

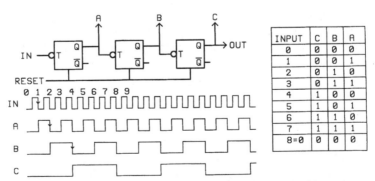

INPUT	C	B	A
0	0	0	0
1	0	0	1
2	0	1	0
3	0	1	1
4	1	0	0
5	1	0	1
6	1	1	0
7	1	1	1
8=0	0	0	0

■ **8-11** *Notice the symmetry of the output pulses in this counter circuit. A toggle flip-flop is a simple way to obtain extremely symmetrical square waves. This circuit counts upward in binary. Frequency counters for the test bench are up-counters.*

This is significant in high-speed logic because the delays can add up to larger amounts of time and cause errors. Other measures can be used to compensate. Different internal logic and a synchronizing pulse keep everything flipping and flopping on time in more sophisticated synchronous counters.

The truth table in figure 8-11 shows that this circuit actually counts in binary. It is also called an *up-counter* because the number value increases as the count progresses until it rolls over back to 0. Figure 8-12 shows how you can also make a binary *down-counter* by changing a few connections.

Down-counters are often used in phase-locked loops. The reason for this is that, so far, the simple chain of flip-flops can divide the input frequency only by some power of 2. If a divider that can be programmed to divide by any arbitrary integer is placed between the VCO and phase detector (see figure 8-1), a programmable synthesizer is possible.

It is easier to make a programmable divider with a down-counter than an up-counter. An up-counter would start at 0 and count up to whatever value is needed, then reset to 0 and start all over again. To recognize when the desired count is reached would require a lot of complicated logic. It would also need to be fast in order to reset the counter quickly enough.

A programmable divider made from a down-counter is much simpler. The counter is designed so that its outputs can be preset to a desired number. It starts counting down from this number. When a count of 1 is reached, it is preset to the previous number. As figure 8-13 shows, the logic circuitry for detecting the one-count is very simple. A NOR gate can do the job. A counter with four flip-flops can be programmed for a maximum division ratio of 16.

Figure 8-14 puts a programmable divider to use in a PLL. The four lines labeled A through D represent four binary bits of data used to

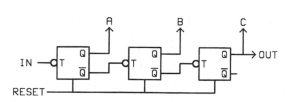

INPUT	C	B	A
8=0	0	0	0
7	1	1	1
6	1	1	0
5	1	0	1
4	1	0	0
3	0	1	1
2	0	1	0
1	0	0	1

■ **8-12** *A down-counter made of toggle flip-flops.*

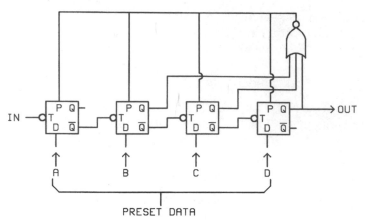

■ **8-13** *This programmable down-counter detects the one-count by using a three-input NOR gate to decide when all but the least significant bit reaches 0.*

preset the counter. In this example, a phase/frequency detector is used. This is a detector which not only provides an error signal useful for detecting phase errors, it also outputs a dc component which changes according to the frequency error.

Both inputs to the detector will be equal in frequency when the loop is in a locked condition. When the divider is programmed to a different division ratio, the VCO is tuned to a new frequency by the error signal. The VCO runs at a frequency equal to the division ratio times the reference frequency.

You could have a system with a reference frequency of 10 kHz and a 4-bit programmable divider like that shown in figure 8-13. Sup-

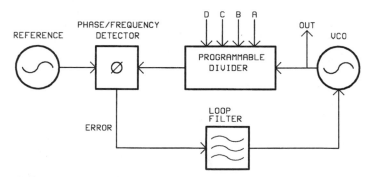

■ **8-14** *Most synthesized radio equipment uses some variation of this type of system. Instead of 16 as a maximum division ratio, real-life synthesizers may have dividers which can count into the thousands.*

pose the division ratio is 15. To operate in a locked condition, the VCO would run at 150 kHz. If you changed the division ratio to 14, the VCO would be moved to 140 kHz.

The phase/frequency detector also uses sequential logic. Figure 8-15 shows a D flip-flop which can be used for this purpose. The name results from its ability to latch onto or remember data present at the D input. The single bit of information at D drives Q to a similar level when the C input is high. When the clock line goes low, this information is retained. The not-Q output level is always opposite of Q.

Level-changes at D have no effect while the clock line is low. When the clock line goes high again, whichever level happens to be present at D will again be sent to the Q output. The D flip-flop can also be fitted with a reset input by adding more logic circuitry.

Figure 8-16 shows how D flip-flops can be used to build a phase/frequency detector. Both D inputs are kept continuously high by supply. The easiest condition to consider might be one in which both frequencies are equal and exactly in phase. The AND gate sees a high at both inputs and outputs a high to both flip-flops, resetting them. In this case, both flip-flops are reset almost as quickly as they are set. Neither flip-flop gets to spend much time in the set condition. Both Q outputs are very short pulses.

If one frequency source leads or lags the other in phase, things are different. The leading positive pulse sets its associated flip-flop. The first flip-flop to become set gets to stay that way until a positive-going pulse from the other frequency source comes along. When this happens, the second flip-flop is set but is almost as

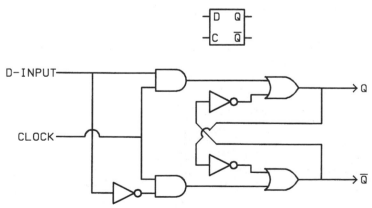

■ **8-15** *This is one way to make a D flip-flop.*

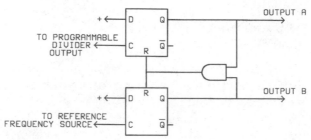

■ **8-16** *A simple phase/frequency detector. Use a NAND gate for D flip-flops which need a low to reset.*

quickly reset by the AND gate. Average output energy from the second flip-flop is almost nil.

As pulses continue to flow, the flip-flop with the leading signal continues to emit wider pulses than the other one. The output pulses can be filtered to obtain an average level for tuning a VCO. If the phase relationship is reversed, the other flip-flop emits wider pulses. If the frequencies are different, the flip-flop receiving the highest-frequency pulse train wins the race most often. It therefore has a higher average output and provides a means of steering the VCO frequency toward lock.

Loop filters

After being filtered, outputs A and B can be applied to opposite ends of a reverse-biased tuning diode, or the differential inputs of an operational amplifier could be used to provide a single-ended output. An even easier method is to dispense with the lower flip-flop's Q output and use the not-Q output. A resistive divider would provide a single-ended output, as shown in figure 8-17.

The loop filter in a phase-locked loop is very important. It can greatly affect how the loop performs. Phase detectors and phase/frequency detectors generate products related to the reference frequency. These must be reduced before being applied to the VCO tuning diode. If the tuning voltage is not sufficiently smoothed by the loop filter, excessive frequency modulation (FM) of the VCO will result.

FM generates sidebands on both sides of the carrier. A 10-kHz reference signal will result in sidebands spaced 10 kHz away from the carrier. A stronger reference component generates more amplitude in the sidebands. FM sidebands theoretically repeat infinitely with decreasing amplitude at the same carrier spacing. This is one

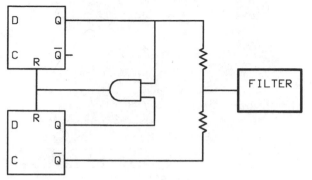

■ **8-17** *An easy way to obtain a single-ended output from the phase/frequency detector.*

reason why poor performance in a PLL can generate a really rotten signal. Figure 8-18 shows how the frequency spectrum around an excessively modulated VCO might look.

Why should filtering a 10-kHz component from the tune line be a problem? If all that we need to do is reduce amplitude of the 10-kHz component as much as possible, plenty of attenuation could be supplied by a multisection filter of some type. One problem with doing so is that too much phase shift may be introduced into the loop.

This is supposed to be a negative-feedback loop. Phase errors should tune the VCO to a frequency and phase angle which produce the smallest error signal. If a phase or frequency error is moving the VCO frequency to reduce error amplitude, the VCO

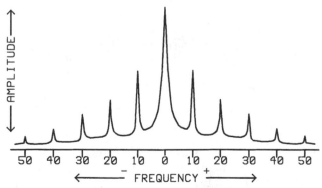

■ **8-18** *This is an example of what the spectrum around a PLL using a 10-kHz reference might look like with insufficient loop filtering. In the extreme case of some early 2-m transceivers, the largest sidebands might be only 30 or 40 dB below the carrier.*

should be able to respond quickly for a number of reasons. Should excessive phase shift in the loop filter delay the corrective change too much, the loop may overshoot.

Excessive phase shift could make the loop overshoot when it tries to correct and tune back in the other direction. It may do this every time it tries to correct. The result is that the loop continuously hunts for zero phase error but never finds it. The whole loop oscillates.

Loop filters usually don't have a lot of sections and end up looking something like figure 8-19. These are low-pass filters. They cause the total loop gain to be higher at lower frequencies and lower for high-frequency changes in the loop. Note that the frequencies mentioned are not that of the VCO output signal.

Looking at the loop as a whole, voltage fed back to the VCO should be opposite or 180° out of phase with any disturbance which would try to force it out of phase lock. To realize a stable PLL, feedback through the loop should not approach 360° at any frequency at which the loop has a voltage gain greater than 1.

Figure 8-20 shows a possible method of measuring voltage gain in the loop. You could break the loop and use an external signal to tune the VCO. A variable dc supply along with a variable-frequency sine-wave generator would be needed. Voltage could be measured at the filter output with an oscilloscope.

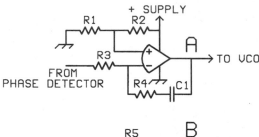

■ **8-19** *R1 and R2 are simply for setting the bias for the op-amp. The A circuit will drive VCOs in a direction opposite of what the B circuit would. Maximum possible phase shift with these single-pole filters is 90°. Be aware that other components in the VCO tuning line may act as low-pass filters and introduce additional phase shift.*

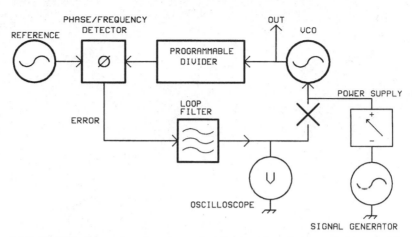

■ 8-20 *This is a setup for measuring the loop's voltage gain along with the patience of the observer. Some systems will be impossible to measure because of difficulties in manually holding the VCO frequency steady and close enough to the reference frequency. A dual-trace scope could be used to simultaneously display the test signal-generator level.*

The external dc supply should be set to a voltage which makes the divider output and reference equal. You could then measure voltage readings of the ac component in the filter output and compare it to the generator reading. The ratio between the two would be the voltage gain of the loop at that particular loop frequency.

This is not always practical because tuning the VCO with a variable voltage can be rather touchy. Voltage gain at very low frequencies is rather high with some systems. A very small change of microvolts in the dc level at the VCO may cause a large change of several volts at the filter output. An experiment like this does at least help illustrate the concept of loop gain.

Practical synthesizer problems and solutions

9

IN THE PREVIOUS CHAPTER, BASIC IDEAS INVOLVED WITH indirect synthesis were presented. This chapter is aimed at explaining and solving some of the practical problems facing hobbyists who wish to build phase-locked loops for HF equipment. Some examples are specifically related to parts used in the multiband transceiver project.

Although not used for the transceiver project, a section of material at the end may be useful if you've ever wished for an alternative signal source. It is a means of providing an accurate frequency readout and excellent frequency stability. It does not rely on special temperature-stable components. It is also simpler than a full-blown microprocessor-controlled synthesizer.

Expanding the frequency range

As mentioned earlier, the transceiver's first local oscillator is a PLL synthesizer operating in a range of 72 to 102 MHz. The example RTL and DTL circuits of chapter 8 are woefully inefficient for working at this frequency range or anywhere near it. Other types of logic do much better.

CMOS integrated circuits are well suited to the tasks involved in a PLL because of their low cost, low power consumption, and medium speed. One chip used in the transceiver contains a few thousand transistors, performs most PLL functions, and operates with input frequencies up to approximately 26 MHz. It contains most of the divider circuitry. The terms *counter* and *divider* can usually be used interchangeably when describing the following circuits.

Other integrated circuits with different characteristics are useful as dividers in the 100-MHz region. A possible PLL arrangement is

129

illustrated in figure 9-1. CMOS is used for all logic except the fixed divider or prescaler. Some TTL chips could be used at the VCO frequency range.

One type of nonsaturating logic where bipolar transistors operate without turning fully off or fully on is called *emitter-coupled logic,* or ECL. The prescaler used in the transceiver project is an ECL chip. ECL dividers are capable of operating at several hundred MHz.

An arrangement like that in figure 9-1 would certainly allow operation in the desired frequency range. The problem with such a scheme is that the smallest possible frequency step is no longer that of the reference. Any change in the divider will be multiplied by the prescaler.

If the prescaler divides by 65, the total division ratio can no longer change by 1. The total division ratio must change by increments of 65 when the main divider is reprogrammed.

Using another programmable divider in place of the fixed prescaler might seem like a sensible solution, but there is an easier way. A fully programmable divider at 100 MHz has been, until recently, impossible and is still impractical. It would be too expensive and power hungry. Dual-modulus prescalers are an alternative solution to providing total divider ratios which can be incremented or decremented by 1. Auxiliary counter and control circuitry is used with dual-modulus counters.

Each time a prescaler counts through and rolls over, the main programmable counter counts down one count. If we want the whole system to change the division ratio by 1, the prescaler must change its division ratio by 1 for only one prescaler-division cycle.

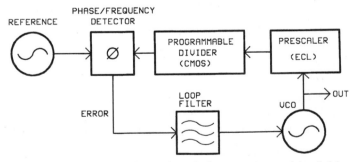

■ **9-1** *A fast fixed divider ahead of a programmable divider allows operation at much higher frequencies than would otherwise be possible.*

This is exactly what happens with a dual-modulus prescaler. The prescaler has a control line which, in the transceiver project, selects a division ratio of either 64 or 65.

Suppose the main divider is programmed to divide by 80, while the prescaler divides by 65. The total division ratio is then 5200. To obtain a total division ratio of 5199, the dividers would need to do the following: instead of dividing by 65 for 80 repetitions, the prescaler should do this 79 times. On the 80th cycle, it should divide by 64. The total division ratio is then $65 \times 79 + 64$, for a total of 5199.

Figure 9-2 shows the connections and overall scheme for the dual-modulus prescaler used in the transceiver project. It's a Motorola MC12017. It has an internal voltage regulator or can accept 5 V at pin 7. One of the nicest features of this chip is the input sensitivity. The minimum drive level necessary for proper operation is approximately 0.5 V peak to peak.

Input and output pins are compatible with standard CMOS. This chip divides by 64 when the modulus-control input is high and by 65 when it is low. Its maximum input frequency is at least 225 MHz.

The main programmable divider is included in a Motorola MC-145158P integrated circuit. This is a CMOS chip, and a partial block diagram is shown in figure 9-3. Two phase detectors provide a

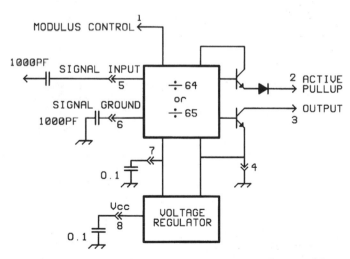

■ **9-2** *Here are the pin connections and an overview of the Motorola MC12017 dual-modulus prescaler. The version I used is packaged as an eight-pin dual in-line package (DIP).*

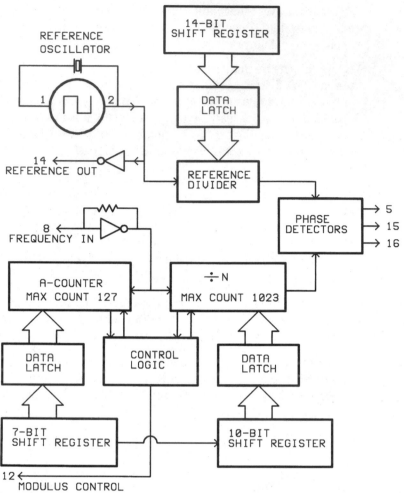

■ **9-3** *This is a partial block diagram of the Motorola MC145158 syn-thesizer chip used in the transceiver project. Although not shown here, the shift registers share common data and clock pins. A selector pin causes data to be sent to one register or the other.*

single-ended output at pin 5 or balanced outputs at pins 15 and 16. The reference frequency is generated by a programmable divider.

If you don't understand what the shift registers and data latches are for, don't worry about them for now. They are simply a means of capturing and storing binary data for the programmable dividers. The MC145158P contains even more functions than are listed in figure 9-3, but we will get to some of them in the next chapter.

The auxiliary or A counter, along with control logic, command the dual-modulus prescaler to change mode at appropriate times. After the A counter counts through its programmed number of cycles, it signals the control logic to change levels at the modulus-control line.

The control logic must also hold the auxiliary count until the main divide-by-N counter finishes. After finishing the main count, both auxiliary and main counters are again preset or reloaded to values in the data latches.

One way to latch a single bit of data is to use a D flip-flop. Any number of bits can be stored by a latch circuit, as shown in figure 9-4. To store more bits, just add more flip-flops. All bits are stored at the same time when the clock line is high. When the clock line is low, output lines retain their status, even though input levels may change.

This is important because shift registers in systems like that shown in figure 9-3 receive their information in a serial format. This means each bit of information is sent over a single wire, one after another. Parallel outputs of the shift register do not end up with the correct levels in one step. In order for the serial-input shift registers to tell one bit from another, they must know when each bit is supposed to occur.

Figure 9-5 is an example of how serial data can be changed to parallel data. Serial-to-parallel converters in ICs normally use more

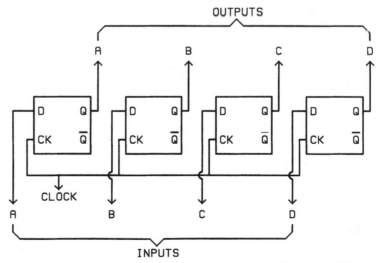

■ **9-4** *A single clock pulse can be used to latch several bits at once.*

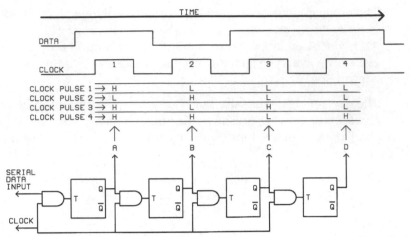

■ 9-5 *This example is a primitive serial-to-parallel converter. Practical converters may use different flip-flops and additional reset inputs.*

sophisticated circuitry than this. Hopefully you will find the principle apparent. In this case, the clock line separates and times the occurrence of each bit. Each clock transition is used to signal the arrival of a new bit.

First, assume all flip-flops are reset with all Q outputs low. The leftmost flip-flop has a high at its input on the first clock pulse. This high is propagated to the A output. All of the other flip-flops have low inputs, thus outputting lows.

On the second clock pulse, things are different. Now, the second, B-output, flip-flop has a high at its input. The first flip-flop input has been presented with a low by the data line. If you look at the levels present at each clock pulse, levels in the data line seem to shift down the line of flip-flops with the occurrence of each new clock pulse.

This example is a 4-bit register but the MC145158P shift registers operate similarly. In the transceiver, a microcontroller generates serial data and clock signals. Firmware used by the microcontroller will calculate the correct binary values to be fed into each of the MC145158P shift registers.

A step at a time

We are well on the way to putting together a PLL that can cover the 72- to 102-MHz frequency range. In order to allow the loop to respond relatively quickly, I picked a reference frequency of

16,384 Hz. I chose 2^{14} because it simplifies some of the calculations performed by the microcontroller.

Tuning the HF bands in such large steps would not be very useful. Instead of the MC145158P on-board oscillator, an external variable-frequency reference oscillator is used in the transceiver project. Many commercial systems have instead included mixing schemes and multiple PLLs to obtain finer tuning steps.

Figure 9-6 is an example of how two loops could be used to obtain 100-Hz tuning steps. Let's use 10 kHz for the reference to make the numbers easier to follow. The output loop now has a mixer and filter inserted before the dividers. A difference frequency from the two VCOs is extracted and filtered before being counted.

Multiple filters would be needed because mixer-sum products will enter the divider unless each bandpass filter is sufficiently narrow. The filters could be selected by a routine in the microprocessor program.

The fine-tune loop on the right fills in the 10-kHz gaps of the coarse tune loop. Its VCO is also locked at 10-kHz increments. After passing through a fixed divider, the fine-tune signal is reduced to changing by 100-Hz steps. The fixed divider also reduces phase noise and reference sidebands of the fine-tune loop by 40 dB.

135

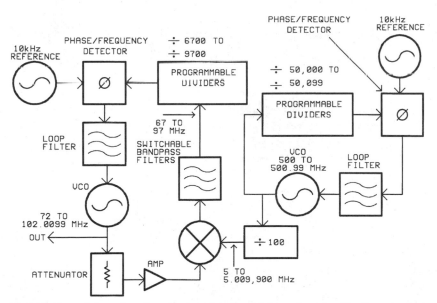

■ **9-6** By mixing the fine-tuning loop's 5-MHz signal with the coarse-tuning VCO, a large 30-MHz span with 100-Hz steps is possible.

This type of system can be expanded to use more loops. Added loops allow finer tuning increments and/or higher reference frequencies. Mixing also generates spurious signals which can end up in the output. This is why an attenuator and amplifier are included in the coarse-tune loop.

A double-balanced mixer will isolate the fine-tune loop from the output. The amount of isolation will probably be only 40 to 50 dB. The attenuator and amplifier can improve that figure. For very high levels of isolation, shielding and decoupling also are important. This includes control lines along with power supply leads.

Using a tunable reference with no mixing in the output loop removes some of the problems of mixing products in the output signal. The actual transceiver reference frequency starts at 16,384 Hz and tunes downward. More information about the reference frequency will be presented in succeeding chapters.

A slow loop

When you add up all of the hardware necessary to make a controller and synthesizer suitable for use in an HF radio, it can look rather intimidating. As mentioned earlier, some of the essential functions carried out by multiloop synthesizers can be accomplished with a simpler method.

Two main requirements of amateur HF operation are frequency stability and an accurate frequency indication. Although programmable operation would be nice, the system I'm about to describe is not capable of retaining frequency information.

This is an easy concept to visualize. It involves using a frequency counter and a VFO. By *frequency counter*, I'm referring to a display-type counter designed to be read by humans, as opposed to counter circuits previously described. The VFO is a conventional LC-tuned oscillator. It could be operated directly at the operating frequency or used in a heterodyne system.

The frequency counter provides a frequency readout, but it can also be used to stabilize the VFO. A human operator could monitor the VFO frequency and correct for temperature-induced drift by retuning any time a discrepancy between actual and intended frequencies is noted. If the person doing this is very diligent, the VFO signal will seem almost as stable as the crystal-controlled timebase oscillator used in the counter.

This may seem to be a silly scheme. After all, would you really want to constantly tweak the VFO while talking to other stations?

If this seems like a tedious task, don't worry—you can easily be replaced by a few components. With this comforting thought of how we humans can be removed from a feedback loop, let's see how the VFO and counter system works.

Display counters

A digital counter displaying radio frequencies is usually designed to count pulses from a source for a specified length of time. If this time window is one second in length, the display units are in cycles per second or hertz. A 0.1-s window results in values with a precision of 10 Hz.

You've seen how flip-flops naturally count in binary. They can count up to a power of 2 and roll over back to 0. It is also possible to make the counter roll over at a different count, such as 10. The result is called a *binary-coded decimal,* or BCD.

BCD counters have four output lines. The largest possible binary number which could be represented by four lines is decimal 15. The largest BCD number allowed is 9. Decoding logic can be used to display conventional decimal numbers.

If you wanted a single output to light a bulb or in some manner to indicate the count of 7, a three-input AND gate, as shown in figure 9-7 would suffice. Any other binary number would be ignored. Other combinations can be decoded by additional logic to produce an output for each decimal number.

Incandescent displays with multiple filaments formed in the shape of decimal numerals were at one time popular. They could be driven with a single line for each decimal numeral.

One of the simplest common forms of decimal display, shown in figure 9-8, is a seven-segment type. Some decoder ICs produce

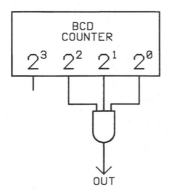

■ **9-7** *An example of how the count of 7 may be decoded.*

137

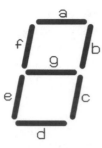

■ **9-8** *LEDs, liquid-crystal displays, and incandescent lamps have all been arranged to form seven-segment displays.*

outputs for them. Each segment has a separate pin. The other side of each light-emitting diode (LED) connects to a common pin. Displays are available with common anode or common cathode connections.

As an example, all the segments in a seven-segment display must be energized to display the decimal numeral 8. Some display-driver ICs require a resistor between each output pin and display segment. Others cycle their outputs on and off to obtain the correct average current level for LED segments.

To display numbers larger than 9, simply add more digits. We do this by chaining BCD counters together in figure 9-9. Each BCD counter is designed to output a single carry pulse when it rolls over. The block diagram represents three of the five digits in a counter which I used to implement this stabilization scheme.

A time-base generator gates the incoming signal on and off. As stated earlier, sending a high to the AND gate for 1 s lets the

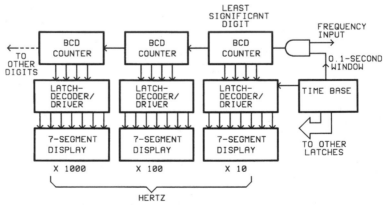

■ **9-9** *A time base in this frequency-counter diagram usually consists of a crystal oscillator, digital dividers, and other logic. It generates the counting window and sends a pulse to the latches after the window has closed.*

counter resolve frequencies to the nearest hertz. While the output is low and the count is stopped, a command is sent to each 4-bit latch to capture and hold the present information.

All that's left is to decode the 4-bit information and drive the display digits. The block diagram of figure 9-9 also represents the counter in one of my receivers. It uses seven-segment LED displays. Some older stand-alone or test-bench counters use similar circuits.

Using the counter's output

A few simple steps are necessary to feed information from the counter back to the VFO. First, the circuit must have some method of deciding if the display indicates that frequency is too high or too low.

We could build a sophisticated circuit which would analyze the counter outputs and compare them with a number programmed in by the operator. Unfortunately, this would be even more complicated than a conventional PLL and response to the programmed values would be too slow.

All that is really necessary for correcting frequency drift is to find out if the VFO frequency is increasing or decreasing. This can be accomplished by testing the output of the least significant digit.

In the system I used, the least significant digit displays frequency to the nearest 10 Hz. The VFO is tuned manually, like any conventional LC-tuned VFO. This is possible because the correcting loop is so slow that it cannot keep up with normal manual tuning rates.

Suppose you've been tuning across the band and then stop. Let's also imagine that the chosen frequency happens to result in a 4 being displayed in the tens-of-hertz (0.01-kHz) position. If, after setting on frequency for awhile, the least significant digit changes from 4 to 5, we know the frequency increased. A very simple decoder can be used to decide the same thing.

This yes/no output of the decoder is sent to a low-pass filter, which greatly slows the voltage change. The slowly varying voltage is sent to a varactor diode in the VFO. As far as the loop is concerned, the VFO is being used as a VCO. The circuit shown in figure 9-10 is capable of tuning its associated VFO over a range of only 6 kHz or so.

Before the corrective loop was applied, the VFO drifted over 2 kHz under some conditions. With the loop, drift is limited by the

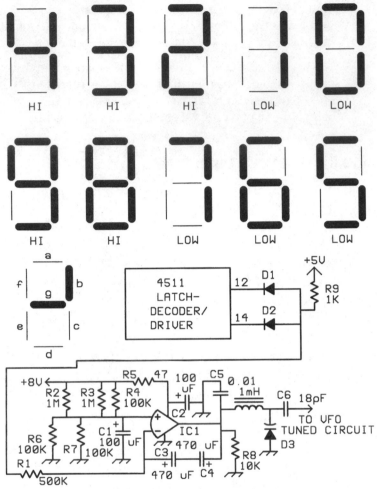

HI HI HI LOW LOW

HI HI LOW LOW LOW

■ **9-10** *Decoding and filtering the output of a frequency counter's seven-segment driver can be used to stabilize a VFO. Pins 12 and 14 of the 4511 drive the b and g segments of an LED seven-segment display. Use a tuning diode at D3 suitable for your own VFO. The same applies for C6. IC1 is half of a TL082 op-amp.*

counter's crystal-controlled time base. This is an easy, no-hassle method of stabilization if you already have a suitable counter from which to derive information.

One limitation of this system is that it is not suitable for rapid switching between two or more frequencies. To use *receive incremental tuning* (RIT) or *transmit incremental tuning* (XIT) in a transceiver, you may want to use other heterodyned oscillators to do the frequency shifting.

Many variations on this idea are possible. Perhaps if we examine figure 9-10 more closely, you will be better prepared to develop your own stabilization systems.

The decoding circuit is made of diodes D1, D2, and R1. It is a diode-AND-gate. Both pins 12 and 14 of the 4511 must be high before the output at R1 goes high. The reason for connecting to the b and g segments is that the numerals in which these segments are used simultaneously are evenly distributed within the total possible number of numerals.

An entire counter is not necessary to make this kind of loop. In the past, other schemes were designed which used a time base and binary counters. Having a display-type frequency counter is much more convenient for the operator.

When operating, the VFO frequency is constantly being pulled up and down. If the change is slow enough, this wobbling effect will be practically undetectable. Capacitor C6 should be as small as possible, while allowing sufficient correction.

The previous comments apply to VFOs, which have sufficient short-term stability to perform well without the loop. Rapid frequency changes caused by external influences must be eliminated or this type of loop will be defeated.

141

Even with ordinary components, you can still build very-high-stability VFOs using this technique. The VFO must employ solid mechanical construction and good electrical isolation. These requirements can be met by careful design and construction. You will not, however, be as dependent on special hard-to-find temperature-compensating components.

A synthesizer for your transceiver

10

SOME OF THE HARDWARE OF THE SYNTHESIZER'S OUTPUT loop has been described. As mentioned earlier, the reference for this loop is varied to provide frequency coverage between the points not otherwise available with only a fixed reference.

A two-loop scheme is used to generate the synthesizer output. Instead of mixing in a fast-tune loop, the reference oscillator is tuned. Tuning the reference entails more number crunching to generate the final frequency. Multiplication and division enter the picture.

This requires that the software is slower and more complicated than the simple addition and subtraction needed in mixing schemes like that shown in figure 9-6. Nevertheless, the 80C31 microcontroller is more than capable of handling the necessary calculations.

A more serious problem, however, is the fact that tuning steps must be so small. Before solving that problem, let's take a look at how the reference frequency is generated.

A variable reference frequency

Figure 10-1 shows a variable crystal oscillator built around Q1, which feeds a buffer stage including Q2. Tuning is accomplished by applying a dc voltage to D1. R1, L1, and L2 form the path which allows this to happen.

The connection at the top of R1 labeled "Tune from DAC" indicates the output of a digital-to-analog converter. The DAC connected to this line is capable of outputting 4095 graduated voltage steps.

Capacitor C5 is useful for trimming the frequency of oscillation. If this oscillator does not cover the correct frequency range,

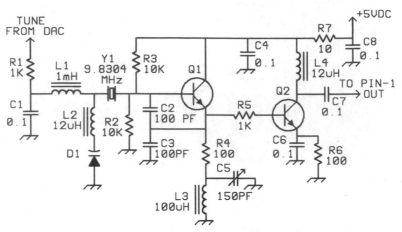

■ **10-1** *This is a schematic diagram of the voltage-tuned, variable-frequency, crystal-controlled oscillator. D1 is a 1N4007 silicon rectifier. Q1 and Q2 are 2N2222A or similar transistors.*

increase or decrease the value of L2. Such steps probably won't be necessary, but keep them in mind.

Starting out at 9.8304 MHz, this frequency is divided by 600 to arrive at 16,384 Hz when at the upper end of the VXO tuning range. The 16,384-Hz reference is multiplied by a value programmed into the output loop.

Although the multiplication factor changes with frequency, changes in the reference are multiplied approximately 10 times at the output frequency. As an example, the output frequency is 98.304 MHz with a value of 6000 in the dividers of the output loop. In this case, a 1-Hz change at the reference results in a 10-Hz change at the output.

A wider reference tuning range is required at the low end of the synthesizer range than near the top end. This is handled by having the firmware figure out how far to tune before a new 16,384-Hz step is needed.

The digital-to-analog converter

To reliably and accurately generate 4095 evenly spaced voltage steps requires a fairly sophisticated circuit. Fortunately, ICs are available to make this part of the synthesizer easier to build.

Figure 10-2 shows the main functions and pin connections of an AD7543 or PM-7543 DAC. This chip uses an *R*-2*R* ladder network similar to the example in figure 6-7. Some digital inputs are gated in a manner which allows much flexibility in how information is used by the serial-to-parallel and latch registers.

Strobe inputs are labeled STB1, STB2, not-STB3, and STB4. They are like the clock input of figure 9-5. One reason for all the extra clock inputs is to allow data to be clocked in on the rising or falling edge of the clock signal.

Not-LD1 and not-LD2 load contents of the serial-to-parallel register into the DAC. Both pins must be low to load the information.

Not-CLR at pin 13, when low, forces the DAC register to all zeros, regardless of the status of any other input pin. SRI at pin 7 is the serial data input. AGND stands for analog ground, which in this application is separated from the main digital ground.

V_{ref} is the reference-voltage input to the ladder network. In the synthesizer it will be connected to ground, although it may be connected to other potentials in other applications.

I_{out1} and I_{out2} connect to the inverting and noninverting inputs of an op-amp. R_{fb} connects to the op-amp output for linear DAC applications. It may be connected differently to perform analog division.

To eliminate the requirement of dual supply voltages for the accompanying op-amp, I_{out2} and AGND are held at a positive potential. This is accomplished in the synthesizer by using the output of a voltage follower made from another op-amp.

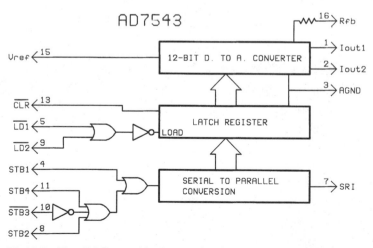

■ **10-2** *The DAC used in the synthesizer is arranged internally, as shown in this block diagram.*

Many other DACs are available. A parallel-input DAC could have been used. Although parallel-input versions can be updated very quickly, serial input allows us to use fewer paths and connections on the synthesizer board.

An ultrafast feedback loop

One way to control the reference oscillator of figure 10-1 would be simply to tune it with a DAC which outputs levels corresponding to known frequency changes. Unfortunately, the accuracy needed in the voltage-to-frequency transfer function would be rather critical. This could be a problem for individuals trying to repeat the same results at home.

At least one commercial manufacturer has an HF transceiver with such a system. By carefully controlling the VXO's response to DAC output levels, a simple fine-tuning scheme is attained.

I wanted to use a closed loop not only for repeatability but to provide better display resolution. Using a dead-reckoning type of system becomes much more difficult when you try to use finer tuning steps along with the ability to display finer frequency indications.

Because of a closed fine-tune loop, the system in this book can resolve and display frequency increments of about 10 Hz. This means the reference oscillator must tune in increments of about 1 Hz. As mentioned earlier, a DDS would have no problem achieving this result and could change frequencies very quickly.

After considering DDS, its expense, and possible output spurs, I sought a different solution. A conventional PLL with a 1-Hz reference was out of the question because it would be far too slow in responding to tuning changes.

Fortunately, there is a way out of this dilemma. Some of the 80C31 microcontroller resources can be used to implement a surprisingly responsive loop. Figure 10-3 helps explain how this is accomplished.

The coarse-tune loop on the left is a conventional PLL, as explained in the previous chapter. The wide arrows indicate information coming from the 80C31 via clock, enable, and data lines.

The fine-tune loop controls the reference VXO. To extract information about the VXO frequency, its signal is mixed with a signal from the central processing unit (CPU) master clock. The single-transistor mixer output is filtered and amplified.

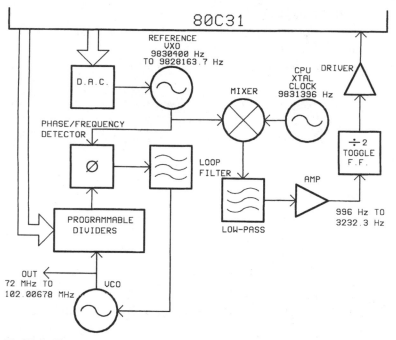

■ 10-3 *The transceiver's synthesizer uses two loops. A conventional PLL is used in conjunction with a frequency-locked loop. The second loop is used to generate a reference frequency. Hardware and firmware routines running on the 80C31 measure and control the VXO to form the second loop.*

A divide-by-two stage provides square waves with a 50 percent duty cycle to the microcontroller. This is actually a D flip-flop wired as a toggle flip-flop. The resulting frequency is low enough to allow hardware on board the microcontroller to measure the period of the square wave.

Firmware running on the microcontroller converts the period measurement to a frequency value. This is much faster than measuring the frequency directly because the circuit does not have to wait an entire second to count how many pulses occurred in one second.

This is why we need a consistent 50 percent duty cycle. Variations in the time an analog signal spends above a trigger threshold could easily result in an unpredictable duty cycle.

Subsequent to calculating the actual mixer output frequency, firmware routines also calculate the target frequency from information displayed by an LCD panel. A value proportional to the difference between actual and intended frequencies is sent to the DAC to correct the VXO.

If the next measurement indicates too much error, another value proportional to the error is sent to the DAC. When the measured error is small enough, the DAC value is left undisturbed until the next error is detected.

The lower limit shown for the reference VXO occurs at the lower end of the output VCO's tuning range. This is where the reference frequency multiplication is smallest in the output loop. As the output loop is tuned higher in frequency, the low-end limit of the reference tuning range becomes higher.

The synthesizer board

Figure 10-4 is a schematic containing most of the circuits involved in the synthesizer. The VXO has already been described. Information about the output VCO will be covered later. Supply regulators and microprocessors are located on other boards.

Components involved in the output loop are at the left of figure 10-4. IC4 is the dual-modulus prescaler. L5 attenuates the VCO signal

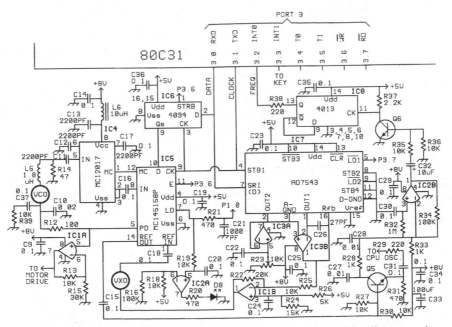

■ **10-4** *Most of the synthesizer circuits are depicted in this diagram. Any standard LED can be used for D8, so pick the style and color of your choice. IC1, IC2, and IC3 are TLC272 op-amps. Q5 and Q6 are 2N2222A or similar transistors. All resistors are ¼ W, 5 percent tolerance.*

before it reaches IC4. It is connected to a 12-in length of RG-158U miniature 50-Ω coaxial cable.

Possibly because of reactance at the input of IC4, a resistive attenuator did not work as well. Some areas of the VCO tuning range were fine, while IC4 miscounted in others. Experimentation may be necessary if you use different terminations or cable lengths.

The remainder of the output loop is concentrated mostly in IC5. Pin 5 connects to the loop filter formed by C10 and R12. It also goes to IC1A, which will be explained later.

Another output from IC4 is pin 7, the lock-detect pin. It emits a series of low-going pulses when out of phase lock. The width of the pulses changes with the amount of error seen by the phase/frequency detector.

When the loop is locked, the pulses are narrow and the output is essentially high. C20 and R19 integrate the pulses into a more constant potential at pin 5 of IC2A. C21 does not need to be connected to port 1.0 unless a simplified version of the synthesizer is built. This will be explained later.

A voltage divider formed by resistors R18 and R16 holds pin 6 at 91 percent of the supply voltage or about 4.55 V. Any time pin 5 exceeds this potential, output pin 7 goes high to light LED D8.

Pin 11 of the MC145158P is the latch-enable line. It allows information to be received over the clock and data lines when it's low. When high, it latches this information into the on-chip registers and ignores any changes taking place at the clock and data inputs.

Pin 1, the reference-in connection receives the output signal of the VXO. The gates used for an on-chip oscillator and reference output come in handy at this point. They provide extra isolation between mixer transistor Q1 and the VXO before sending the signal out on pin 14.

C15 passes the reference-oscillator signal to the base of mixer Q5. The CPU clock oscillator is injected into the mixer circuit at R29. C27, R28, and C28 form a low-pass filter to help remove RF components before the difference frequency is amplified by IC2B.

R32 and C30 restrain pin 3 from making any fast voltage changes. Pin 2 is unencumbered by such filtering and is the input of an inverting amplifier. Although pin 2 and pin 3 are at about the same dc potential, ac potentials differ when a signal is applied.

An amplified version of the mixer output is coupled through C32 and R35 to Q6. Pull-up resistor R37 allows Q6 to drive a clock input of IC8. After IC8 does its job of halving the mixer frequency, output pin 13 sends the signal through R38 to port 3.2 of the 80C31 microcontroller.

Although it may not be absolutely necessary, here is the reason for R38. Pins of port 3 on the 80C31 can act as outputs as well as inputs. Under some conditions, the pins of port 3 could output a high or low, while IC8 does the opposite.

A direct connection under such conditions would not be healthy for either chip. A power surge or a chaotic power-up sequence might cause port pins to do the wrong thing. If this happens, R28 can dissipate some power instead of the chips having to do it all.

Current in the R-$2R$ ladder network of the DAC is converted to a corresponding potential by op-amp IC3B. Instead of a separate negative supply for V_{ref}, another op-amp is used to establish a potential between V_{ref} and analog ground.

IC3A establishes a "stiff" voltage level at output 2 and analog-ground pins on the DAC. Its output voltage is established by R23 and should set to about 4 V.

IC1B increases the range of voltage swing at IC3B. It also reverses the direction in which voltage changes occur.

This takes care of most of the synthesizer circuits. IC6 is involved only because the serial data stream passes through it to get to IC5. Output pins on IC6 drive the control board and band-switched filter boards described earlier in the book.

Wiring the synthesizer board

Locations of components on the synthesizer board are shown in figure 10-5. Most of the wire jumpers are drawn in as long meandering lines between pads. Some of the power connections which need jumper wires are simply labeled.

Various types of multipin connectors and ribbon cable are available to match the 0.1-in spacing at connections P3.0 through P3.7. The same is true for making connections at the IC6 interface.

Inductors L7 through L14 are 100-μH chokes. C37 through C44 are 0.1-μF ceramic capacitors. These components could be mounted off-board at a cable-entry point. Their purpose is to keep switching noise out of the radio circuits.

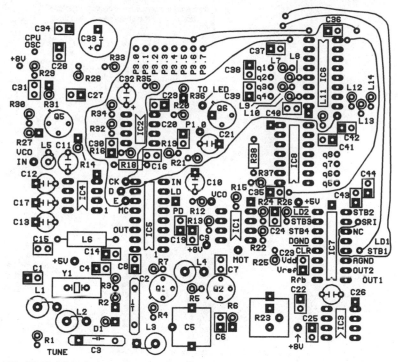

■ 10-5 *This is how the synthesizer-board parts are laid out. Jumper wires are drawn for clarity but do not have to be routed exactly as shown. C37 and R39 are not shown here. They should be installed at the junction of C10 and R12 to improve performance of the loop.*

It goes without saying that IC sockets are a good idea on a board of this complexity. Even IC4 should be in a socket, although it should be a low-profile type to avoid excess parasitic inductance.

Be careful with the orientation of ICs and sockets. Note that they are not all pointing in the same direction. TO-39 metal-case transistors are shown, but plastic-case types will work as well if they are inserted correctly.

As usual, square pads represent locations where leads should be soldered to the ground plane. Connections to the VCO output and tuning input should be made with miniature coaxial cable.

When mounting the board, don't let anything short out paths on the board. This board and other circuits directly connected to the microprocessor should not be mounted near any of the previously described radio circuits.

A ground-plane view of the synthesizer board's etching pattern is shown in figure 10-6. A view of the etched side is shown in figure 10-7.

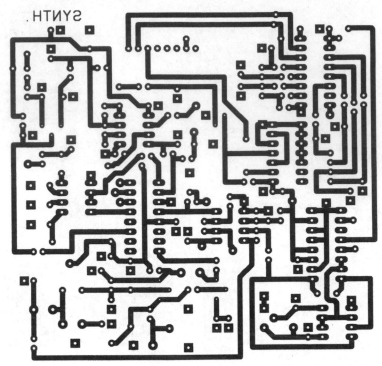

■ **10-6** *This a view of the etching pattern looking through the ground plane and substrate (actual size).*

Early testing

To completely check out the synthesizer, all of the remaining transceiver hardware must be in place and operational. Some of the synthesizer circuits can be tested before you get to that point. This is a good idea because other circuits may be more easily brought on line if these are known to be functional.

To begin the test, plug in ICs 1, 2, 3, and 8. Leave ICs 4, 5, 6, and 7 out of their sockets. Adjust R23 to midrange or 5 kΩ from wiper to ground. Temporarily jumper pins 1 and 2 of IC3B together.

Apply regulated 5- and 8-V positive supplies to the appropriate board connections. Connect a variable-frequency signal generator to R29. Set the output level to about 2 V peak and set the frequency to approximately 9.83 MHz.

Look at the VXO output at C7 with a frequency counter and/or oscilloscope. Check the signal and measure its frequency to see if it is close to 9.83 MHz.

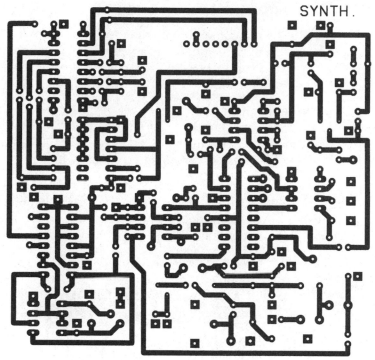

SYNTH.

■ **10-7** *This is a direct view of the synthesizer etching pattern (actual size).*

If the VXO is working, check the dc voltage on the op-amp end of R1. While you are doing this, crank R23 back and forth to see if you can adjust the VXO to exactly 9.830400 MHz.

Next, see if you can adjust R23 to continuously lower the frequency to at least 9.828 MHz. If not, adjust C5 and try again. If the VXO responds correctly, proceed with testing the mixer.

With an oscilloscope hooked to the collector of Q6, you should be able to see a waveform. The frequency of this signal should be much lower than in the first test, so you should readjust the time-base setting on the scope.

The frequency of the signal at Q6 should change as you tune the signal generator. While you are doing this, leave the generator frequency at some convenient setting and check the waveform at pin 13 of IC8. It should be a square wave with a frequency exactly half that of the Q6 signal.

This is about all of the hardware which can be tested at this stage of construction. The other chips depend on the microcontroller for instructions. Don't forget to remove the temporary jumper on IC3.

The 80C31 microcontroller as a counter

The 80C31 has some on-chip timer/counter hardware which can perform many functions while various computing chores are occurring. Figure 10-8 is a symbolic representation of how a timer/counter section operates.

In the transceiver project, we will be using the timer function. That may sound contradictory, but remember that in this application the period is measured and converted to a frequency value. It's an indirect way of obtaining a frequency count.

To explain why this roundabout method is used, suppose we need to measure the frequency of a signal which could be in the area of 1000 Hz. A conventional counter must count for a full second before you can find out if 1000 pulses have occurred.

If you have an accurate method for measuring the period of this waveform, only one cycle is needed to determine the frequency. For the 1000-Hz example, 0.001 s is all that is needed, assuming a constant frequency in both cases.

Computing the frequency requires some additional time, but the microcontroller can do this much more quickly than would be

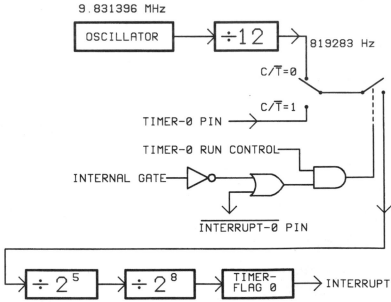

■ **10-8** *This is not a literal interpretation of the 80C31 timer/counter, but it illustrates how various software and hardware inputs can exert control to configure it as needed.*

required to count the other 999 pulses in this example. Frequency is the reciprocal of the period, or put another way:

$$\text{Frequency} = \frac{1}{\text{period}} \qquad (10\text{-}1)$$

The on-chip timer, in this case timer 0, counts pulses which occur at $\frac{1}{12}$ the CPU clock frequency. A command in firmware sets the timer/counter selection to operate from this divide-by-12 stage instead of counting pulses from the external timer-0 pin.

The timer-0 run control is also a firmware command. In this case, it is set high, as is the internal gate command. Under these conditions the "path" to timer 0 is ready to be closed and allow counting to commence. For this to happen, the interrupt-0 line connected to port 3, pin 2 (port 3.2) must be high.

To obtain a frequency measurement in hertz, the divide-by-12 stage output frequency is the defining unit. Firmware routines could divide 819,283 by the timer count. This would give the mixer frequency in hertz.

The only problem is that resolution is not high enough for the shorter periods. If the divide-by-12 stage could output a much higher frequency, the period measurement could be accomplished with greater precision. Unfortunately, hardware limitations place a limit on how far this tactic will get us.

Another way to resolve smaller frequency changes is to measure several periods before calculating the frequency. This approach requires a little more time, but it's still faster than counting pulses directly.

Here is how the microcontroller determines the mixer frequency: when P3.2 (interrupt 0 pin) is high, counting commences. It continues until IC8 drives P3.2 low. At this point, several things happen.

The pulse counting stops. Other hardware inside the microcontroller senses the high-to-low transition and allows whatever the program is doing at the moment to be interrupted in an orderly manner.

The interruption is used to evaluate whether enough pulses have been counted to make a calculation. If so, the number of pulses is temporarily saved along with how many periods were counted. The counter is then reset to 0 and the program returns to whatever it was doing before being interrupted.

Now the program has all the information it needs to calculate the mixer output frequency. If written in the form of an equation, it looks like this:

$$\frac{\text{Timer-cycle frequency} \times \text{number of periods}}{\text{timer count}}$$

$$= \text{mixer-out frequency} \quad (10\text{-}2)$$

The grouping of terms may not be important algebraically, but it affects the precision of the program's calculation. If timer-cycle frequency is divided by timer count before dividing by number of periods, rounding errors occur because of remainders in the quotient. Alternatively, more complicated and slower math routines would be required in the firmware.

To complete the explanation of figure 10-8, the timer flag can be affected when the counters roll over. A program interrupt can also be generated at rollover. These features are not used in this application.

Completing
the output loop

11

AS MENTIONED PREVIOUSLY, A VCO WITH A LIMITED TUNING range eases many of the problems encountered in designing a well-behaved synthesizer. An oscillator which covers 2 or 3 percent of the output frequency is much easier to deal with than one with a 30 percent range.

The problem we have here is that the synthesizer is designed for a range of approximately 72 to 102 MHz. For 2 percent coverage, we need 15 separate ranges. This is certainly feasible but requires a certain amount of space.

There is good reason to narrow the range even more if possible. The reason relates to the digital phase detector in the MC145158 chip and most other systems, too. Pulse-width modulation is used to vary the phase-detector output.

High- and low-going pulses at the single-ended phase-detector output are very narrow when the filtered output is centered between supply and ground. Being so narrow, they have relatively little energy and are more easily attenuated by the loop filter.

If the loop needs to move the output frequency above or below this point, the dc component also moves. Pulse widths of one polarity or the other become wider and contain more energy. More of the pulse energy rides through the filter and causes larger reference sidebands.

It would be nice if we could narrow the tuning range even more to maybe 100 kHz and have each one selected automatically. To build such a system, we would need 300 separate voltage-tunable ranges. This is the point in the project at which you might want to consider purchasing a good stereo microscope.

OK, I'm just kidding about the 300 tuning ranges and microscopic components. There is, however, a way to accomplish the same result without extreme miniaturization.

VCO and amplifiers

Figure 11-1 shows a more practical approach to building the VCO and buffer amplifiers. A small, mechanical variable capacitor is used to achieve a 30+-MHz tuning range. Electronic tuning is provided by D1. IC1 and IC2 are monolithic integrated circuits which sample a small portion of the VCO energy and build it to a level suitable for the first mixer.

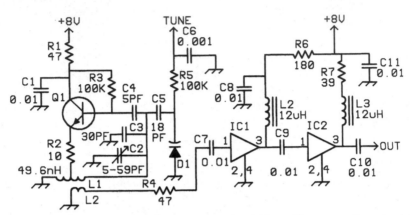

■ **11-1** *C2 is a Mouser Electronics polyethylene-film capacitor, stock number 24TR222. Mounting and shaft screws are stock number 48SS003. D1 is an MV2104. IC1 is a Mini-Circuits MAR-1 marked by a dot on pin 1. Pin 3 is 180° opposite. IC2 is a MAV-11 from Mini-Circuits with a dot marking pin 3. L1 is made of four turns of AWG number-18 to -22 wire wound on a ¼-in form spaced about eight turns per inch. Adjust spacing for the required frequency coverage. L2 consists of nothing more than stray inductance of the leads of R4 and C7. Q1 is a BFR90. Use a ferrite bead on the base if necessary.*

C2 can be controlled manually or automatically. The simplest manual control method is to set the displayed frequency, then turn C2's knob to the setting where phase-lock occurs. Another semi-automatic method will be described later, as well as a fully automatic method of tuning C2.

Transistor Q1 is biased only by R3 to reduce loading of the tuned circuit. This may not be the best way to build the oscillator, and a conventional H-bias scheme may be more reliable. To do this, add a 10-kΩ resistor from base to ground and change R3 to 15 kΩ.

Another possible place for improvement is to substitute an RF choke for R5 and make C5 smaller. Figure 11-1 represents the

VCO at a point where the transceiver is operating well, and I don't find any evidence of excessive phase noise.

With so few components in the circuit, it would be easy to simply solder them to an unetched board at ground points and connect the other leads together at the appropriate points. If you prefer a printed circuit pattern, a layout is shown in figure 11-2.

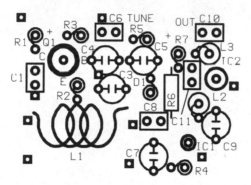

■ **11-2** *Traces for this board are on the same side as the components. Don't drill holes for the component leads. Simply solder them to the traces. When tuning L1, the hot end of L1 may tear loose unless you press down against it with a tool during the process.*

Inductor L1 is adjusted to the correct value by expanding or compressing the length of the coil. C2 is not shown because it is mounted separately on a box wall. Leads for C2 should be as short as possible.

Resistors are easier to fit into the layout if they are ⅛-W types. A ¼-W unit must be used at R7. This layout is used for surface-mounting. The only place you need to drill holes for wire leads is at the square pads.

You can make "feet" for conventional wire-lead components by bending the leads. Real surface-mount components can also be used in some of the locations.

The transistor and integrated circuits have round, pill-shaped bodies. To mount them, drill a hole slightly larger than the body. This will allow the leads to seat flat against the board.

With the components and traces on the same side, the opposite side of the board serves as a shield and ground plane. Insert bare wire through each hole marked by a square pad and solder on both ends.

The suggested etching pattern is located in figure 11-3. It is sized to fit against the opening of a small, stamped-metal box available from Mouser Electronics, stock number 537-M10-PLTD.

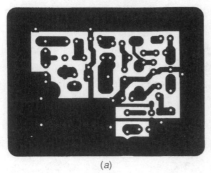 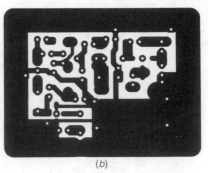

(a)   (b)

■ **11-3** *(a) A direct view of the board paths. (b) The view looking through the board substrate (actual size). The wide copper area around the periphery is used to contact the box opening, or the entire board could be mounted inside a larger box.*

VCO mechanical ideas

To position capacitor C2, a small motor and homemade reduction drive are used. Figure 11-4 shows how the motor shaft bears against a drive wheel. The drive wheel is bolted to the shaft of C2.

The VCO enclosure doubles as a support for these mechanical components. Notches cut into edges of the box allow passage of supply, tuning, and output wires.

Brass stock (0.010-in) is wrapped around the motor and soldered to the box walls. The white material covering part of the motor is foam packing tape for cushioning.

In addition to forming the sheet brass around the motor, four holes were drilled in the brass. AWG number-18 copper wire was inserted and twisted to cinch the brass tight against the motor body.

Figure 11-5 shows how the brass continues as an arm over to a stud and nut. The stud is made of a 4-40 screw soldered to the box wall. Turning the nut allows you to adjust how much pressure is exerted against the drive wheel.

The drive wheel is a 3-in-diameter plastic lid from a peanut butter jar. After marking and drilling the center, the lid is mounted on a drill press and turned against a wood chisel to cut a shallow channel in the outside edge. An ordinary elastic rubber band is then strung around the outside edge to form the drive surface.

Although the drive system is somewhat crude and definitely inexpensive, it has been very reliable. Some builders may want to try

■ **11-4** *A view of the completed motor-driven VCO. Machine nuts are soldered inside the metal enclosure. Matching screws are used to pull the circuit board tight against the box.*

■ **11-5** *The motor shaft is pressed against the drive wheel by tightening a threaded nut.*

using a high-quality capacitor and gear motor for the VCO. Such a system should be even more reliable and less susceptible to microphonics.

I used a Radio Shack motor, part number 273-223. I estimate the rotational speed of the drive wheel to be about 2 r/s. This is probably the maximum practical speed. Anything faster may cause control problems.

If you use the Mouser capacitor or another with a rotation stop, make sure the drive wheel is secured to the capacitor shaft with a star washer. When the transceiver first powers up, the synthesizer has not yet received its instructions and the drive wheel rotates and jams against the stop.

A variable capacitor with no stops would be advantageous in this case. A variety of small air-variable capacitors are available, although some of them are rather stiff and may require too much torque from a small motor. The best candidates would be ball-bearing types.

Speaking of motors, the one in the prototype seems to have more than enough torque. As a result, it is a bit power hungry. Depending on your situation, a lower current motor may do a good job. One candidate is another Radio Shack (special purchase) part, number 273-237. They have small pressed-on gears which can be pulled off to expose the shaft.

Another possibility for mechanical drive components exists in replacement parts made for tape recorders, VCRs, and office equipment. You may be able to adapt some of the rubber-edged, friction-driven wheels instead of fabricating them.

Figure 11-6 is a photo of the VCO and motor drive nestled into the prototype transceiver. The motor control transistors are at the upper right. The synthesizer board is at the upper left, above and adjacent to the VCO. This compartment also contains the display, three-chip computer, tuning mechanism, and voltage regulators.

■ **11-6** *A view of the VCO shoehorned into the synthesizer compartment. Whether you choose smooth or crunchy, most peanut-butter jar lids have a small mold mark underneath, which is precisely in the center.*

Unfortunately, I left insufficient room for a good cushioned mount to isolate the VCO against shock and vibration. Some cushioning material is present, but it is, of necessity, rather thin. Small coil springs mounted on wire arms help hold the VCO in place by pressing upward against the compartment cover.

Motor controller

It isn't hard to see how a small permanent-magnet motor can turn a capacitor shaft. A more puzzling question might be "How does it know where to turn the shaft and when to stop"? Part of the answer lies in figure 11-7.

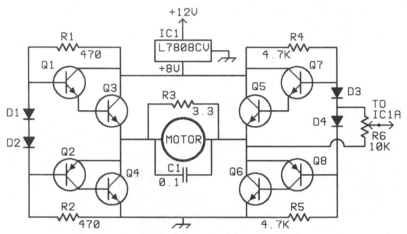

■ **11-7** *This active bridge drives a small permanent-magnet motor to coarse-tune the VCO. Q1 and Q7 are 2N2222A or similar transistors. Q2 and Q8 are 2N2907A or similar. Q3 through Q6 are 2N5491 or similar transistors. For diodes, use 1N4148 or similar silicon types.*

This is a type of active bridge circuit which drives the motor when it's unbalanced. The left side of the circuit, including Q1, Q2, Q3, and Q4, is biased to keep the motor lead at 4 V. If the right side, including Q5, Q6, Q7, and Q8, receives a 4-V input at R6, the bridge is balanced and the motor is stationary.

Using active components for all four legs of the bridge eliminates heavy standing currents which would be necessary to turn the motor with a resistor bridge. When this bridge is balanced, only a few milliamperes are drawn by the idling transistors.

IC1A in figure 10-4 has a voltage gain of 1.6 so that the synthesizer's phase-detector output must be 2.5 V to balance the bridge. When it deviates enough from this value, the motor rotates tuning capacitor C2, shown in figure 11-1, to reduce the unbalanced condition.

Capacitor C1 is used to reduce electrical noise generated by motor brushes and commutator. The purpose of R3 and R6 is to reduce a mechanical oscillation sometimes referred to as *hunting*.

As the motor rotates C2 to balance the bridge, it eventually reaches that point and motor current drops to zero. Inertia of the motor armature allows it to continue spinning.

Feedback from the phase/frequency detector corrects the overrun error, sending the drive wheel in the opposite direction. If rotation is reversed too forcefully, it will continue past the balance point and repeat this back-and-forth sequence indefinitely.

To allow the system to stabilize, R6 must be adjusted to a suitable value. Mechanical and electrical losses must be high enough to dampen overruns. R6 affects system gain so that you can have enough for reliable motor operation but not so much as to overwhelm damping losses.

The purpose of R3 is to provide some intentional electrical loss. A coasting permanent-magnet motor acts as a generator. If the generated power is dissipated in a load, energy stored in the rotating armature mass is removed, slowing the motor.

A different value may be required at R3 if different drive systems or motors are used. By adjusting R6 and varying R3, this control system should work with a wide range of mechanical configurations.

Component arrangement of the controller board is not critical and perforated-board construction is acceptable. Component locations for a printed circuit board are visible in figure 11-8. To fabricate this board, use single-sided stock.

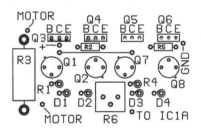

■ **11-8** *Part locations for the motor controller board. A jumper wire on top of the board connects Q1 and Q3 collectors to Q7 and Q5. C1 is not visible because it's soldered directly to the motor.*

Power transistors Q3 through Q6 must be bolted to a heat sink, using insulated hardware. I used a single piece of 0.080-in-thick aluminum with an area of about 8 in^2.

Power transistors can be mounted at right angles or parallel to the board. Mounting the board and heat sink in the same plane is probably easiest. By cutting an aluminum sheet to shape, you can drill mounting holes and bolt them together. Figure 11-9 shows an etching pattern for the board.

■ **11-9** *Use single-sided board stock here. (a) can be used to reproduce the board via photoetching; (b) is a direct view of the etched paths (actual size).*

New ideas

In actual use, the motor-tuned VCO has been a satisfying and valuable section of the transceiver. Most operators don't make large frequency excursions nearly as often as they tune around small segments of the band. Even if you change bands a lot, the running time is very brief, so the motor should last a long time.

Encouraged by the results of this scheme, I can't help but think that someone might develop other ways to remotely change capacitance or inductance. Alternate methods for moving one of the capacitor plates might be employed.

Sometimes materials and processes become available which make new solutions possible. Plastic piezoelectric film could be used to move the plates of a capacitor instead of a motor.

Maybe a small magnetic pump could be used to move iron powder suspended in oil to vary inductance? Could technology similar to the liquid crystal display vary capacitance by changing the orientation of conductive particles?

Even thermal effects can be exploited if power requirements for the heating element are small enough. The best approach would probably be to effect a change which would cease and remain static when heating power is removed.

A thermally powered device may not be as big and awkward as you might think. Sandia National Laboratories in Albuquerque, New Mexico, has developed a steam-powered piston which measures about 6 μm across.

Maybe the market for a better remotely tuned capacitor is too small to justify the development and production of a new device. After all, varactor diodes are widely used in spite of their shortcomings.

Change seems inevitable and a new device could become a catalyst. Oscillator performance is not the critical issue in most consumer equipment that it is in high-performance amateur gear. Front-end filtering for receivers, however, is a different matter.

Many receivers could benefit from the use of front-end tracking filters. As RF signals proliferate, an inexpensive device, immune to strong-signal overload, would be very desirable.

167

It computes!

12

THE CIRCUITS AND HARDWARE DESCRIBED TO THIS POINT perform many useful functions in the transceiver. To control and coordinate various actions, a small digital computer is used. Its resources are minuscule when compared to today's personal or desktop computers. Although it's not the type of system which will automatically make QSOs, fill in the log, and communicate with a smart coffeepot, it adds a level of machine intelligence which makes the transceiver fun to use.

This chapter contains information about the raw material used by the computer: binary numbers. Next, some of the computer hardware is explained. Finally, construction of the single-board computer is described.

169

Binary numbers

Everyone is familiar with the decimal number system and how to count and represent numbers in written form. We have 10 different symbols, or numerals, to represent whole numbers. The value of a numeral depends on where it is located in relation to other numerals representing a number. As we count from zero to nine, one digit is sufficient to represent the numbers. On the tenth count, we write 10 to indicate one unit of tens and no ones.

A similar system is used in binary; however, the highest possible number represented by a single digit is one. If you start at zero and count upward in binary, the count proceeds as 0, 1, using single digits. To count any higher requires another placeholder representing the next-higher power of two. Two in the decimal system is represented as 10 in base-2 notation. Three would appear as 11.

Figure 12-1 shows how the decimal number 15 is represented in base-10 notation, contrasted with how the same number is represented in base-2 notation. It obviously takes more digits to represent binary numbers than the same number would in decimal form. This is true whether the representation is on paper or in the registers made of electronic circuits. Although more circuits are required for binary representation, they are simpler and more reliable than circuits which must function with 10 different voltage levels.

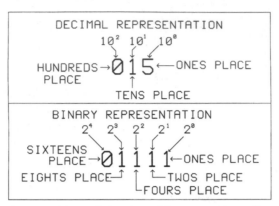

■ **12-1** *Most of the time, binary numbers live deep inside computers and we don't have to deal with them directly. If you plan to modify the program that runs the radio, plenty of them will be available to educate and entertain you.*

The 1s and 0s in binary notation are generally called *bits*. The 80C31 microcontroller is an 8-bit chip. For most arithmetic operations, an 8-bit number is the largest value which can be manipulated in one instruction cycle. To handle larger numbers, software for the 80C31 must be designed to break up the computational chores into intermediate steps until the solution is obtained.

Binary notation is sometimes useful, but it's also rather awkward for humans dealing with large numbers. Other number bases are often used to represent binary numbers in a more manageable format.

The 8-bit numbers handled by the microcontroller are often referred to as *bytes*. A number system sometimes used in computer techniques is the *octal system*. Each numeral represents 8 bits. One advantage of the octal system is the simplicity of converting from binary to octal or octal to binary. The octal system uses the following characters: 0, 1, 2, 3, 4, 5, 6, and 7. An octal value greater than 7 must be expressed by using multiple digits.

Another popular way to view binary information is the *hexadecimal number system*. Hexadecimal, or base 16, uses positional values that are powers of 16. Decimal digits 0 through 9 will represent only 10 values; therefore, we need six additional symbols. The letters A through F are used to represent these additional values. Hex digits 0, 1, 2, 3, 4, 5, 6, 7, 8, 9, A, B, C, D, E, and F correspond to decimal values 0 through 15.

Hexadecimal numbers are also easy to convert to binary. To do so, write a four-digit binary number for each hexadecimal digit. Place the four-digit binary groups in the same order as the hexadecimal digits.

To convert the hexadecimal number 3F to binary, write 0011 for the first hex digit and 1111 for the last one as in 00111111. You can reverse the process by breaking binary numbers into groups of four and converting them to hexadecimal digits.

The octal and hexadecimal systems are not used by the computer; they are a means of condensing binary information so that you don't have to look at long rows of 1s and 0s every time you want to deal with a number. To convert binary, octal, or hexadecimal numbers to decimal, all that's necessary is to multiply each digit by its place value and add the products.

Using the bottom half of figure 12-1 as an example, we could proceed from left to right adding in base 10: $1 + 2 + 4 + 8 = 15$. The same procedure can be used for octal and hexadecimal conversions. Just remember to use the proper exponent for the multiplier of each digit.

Converting decimal numbers to other number bases requires more steps, but there is a straightforward method of conversion. Figure 12-2 breaks down the steps in converting decimal 29 to a binary number.

```
29/2 = 14 R 1
14/2 =  7 R 0
 7/2 =  3 R 1
 3/2 =  1 R 1
 1/2 =  0 R 1
```

■ **12-2** *This is an example of a procedure that converts decimal numbers to binary. It could easily be implemented in a simple computer program.*

We can divide the original decimal number by 2 and check for a remainder. If the remainder is 1, a binary 1 is generated. Repeat this division by 2 until the quotient equals 0. The least significant bit is the first remainder. The most significant bit is equal to the last remainder.

You can also convert other number bases, such as octal or hexadecimal to decimal, in this manner. To convert hexadecimal numbers, divide by 16 at each step and place the remainders in the same order, least significant digit at the rightmost position.

Binary arithmetic

Various software vendors supply computer programs which allow you to convert between number systems or perform arithmetic operations in various number systems. In addition to using personal or desktop computers for this task, some electronic calculators are suitable.

Such products are especially useful with large numbers. At other times you may find the ability to perform manual binary arithmetic useful. This may occur when trying to write some low-level software procedure where insight into the math helps in designing the program.

Arithmetic processes in the binary system follow the same basic rules as in the decimal number system. Just as placing a 0 on the right of a decimal number increases its value by a factor of 10, a binary number is twice as large when similarly shifted left one place. Shifting numbers left or right allows binary computers to perform a very fast multiply or divide operation when the multiplier or divisor is a power of 2.

Normal arithmetic operations can be carried out on numbers in any scale, but the rules in the binary system are extremely simple. For instance, addition has only three rules for single-digit numbers. $0 + 0 = 0$, $1 + 0 = 1$, and $1 + 1 = 2$, expressed as 0 with 1 to carry.

Only three rules are necessary for multiplying single-digit numbers. $0 \times 0 = 0$, $1 \times 0 = 0$, and $1 \times 1 = 1$. The rules for multiplication can be used to perform division. Addition rules can also be used for subtraction.

Figure 12-3 shows an example of addition, subtraction, multiplication, and division with corresponding decimal numbers. The addition example requires carries to the second, third, and fourth digits, reading from right to left.

For subtraction, it may be necessary to borrow from the preceding column. This example borrows from the second, third, and fourth columns.

```
      ADDITION              SUBTRACTION

     00111  =  7₁₀          01100  =  12₁₀
   + 00101  =  5₁₀        + 00111  =  7₁₀
     01100  =  12₁₀         00101  =  5₁₀


   MULTIPLICATION          DIVISION
        111  =  7₁₀            111
     X  101  =  5₁₀      11 ⟌10101    21₁₀/3₁₀ =  7₁₀
        111                   11
        000                   100
        111                    11
     100011  =  35₁₀           011
                               11
                                0
```

■ **12-3** *Binary arithmetic can be performed in a manner similar to ordinary decimal arithmetic.*

Binary multiplication is similar to decimal multiplication. The intermediate addition part of this example also requires carries to the first four digits.

Binary division looks strange because of the intermediate multiply and divide operations which are carried out in the binary system.

Pure binary representation of a number is often not suitable for a display because it is not easily interpreted by humans conditioned to decimal numbers. Pure decimal representation, however, is also unsuitable for large numbers. A display driver with several thousand separate output lines would be ridiculous and unnecessary.

Since the decimal number system relies on combinations of digits to represent numbers, all that's necessary is to encode 4 bits to drive each digit. Such codes are known as *binary coded decimal,* or BCD, codes. An easy way to implement such a code is to use the first 10 binary numbers (0000 through 1001 in binary) and reject the remaining 6 (1010 through 1111) illegitimate codes. Figure 12-4 is an example of how a BCD number would look when applied to a four-digit decimal display.

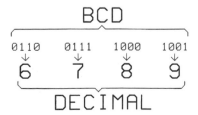

■ **12-4** *Binary coded decimal is often the most convenient way to encode information for driving numerical displays.*

Negative numbers

Computers must have some method to deal with negative numbers as well as positive numbers. This is normally done by using a single bit as a sign indicator. The conventional way of doing this is to make the most significant bit a 1 for negative numbers. A positive number would have a 0 in the most significant bit. This method cuts the largest possible positive value in half, but an equal number of negative numbers can now be represented.

The remaining bits below the most significant bit directly indicate the value of a positive number. Negative numbers could also be represented this way; however, it results in the strange situation of having both positive and negative 0s.

A better method is to use a representation for negative numbers called *twos complement*. The sign bit is still used to indicate whether a number is positive or negative, but the remaining numbers do not directly indicate the magnitude if the number is negative.

To represent a negative number in twos complement form, we first form the ones complement of the number in binary. The ones complement is formed by changing all 1-bits to zero and all 0-bits to one. For example, the ones complement of 01100011 (99 in decimal form) is 10011100.

The twos complement is formed by adding 1 to the ones complement, as in figure 12-5. Instead of being used directly, the resulting number is usually an intermediate step in a calculation involving subtraction or negative numbers. Twos complement numbers have the advantage that they can be added without concern for the sign. No conversion is needed.

$$
\begin{array}{r}
10011100_2 \\
+\ 00000001_2 \\
\hline
10011101_2
\end{array}
$$

■ **12-5** *After converting to the ones complement, the twos complement is formed by adding 1.*

As an example, we should be able to add decimal 89 to −89 and obtain 0 as the result. The binary form of 89 is 01011001. The ones complement is 10100110. Adding 1 to get the twos complement, the result is 10100111. As figure 12-6 shows, adding the two numbers together results in 0 when the carryout of the sign-bit position is ignored.

$$01011001_2 = 89_{10}$$
$$+\ \underline{10100111_2} = -89_{10}$$
$$00000000_2 = 0_{10}$$

■ **12-6** *Adding the twos complement to the original number results in a sum of 0 if the carry is discarded.*

What is a microcontroller?

You might say a microcontroller is an integrated circuit which contains all or almost all of the hardware necessary to run programs, take inputs, and deliver output signals. By contrast, a microprocessor in a personal computer is useful only when connected to external memory and peripheral circuits which communicate with other devices or an operator.

Like most digital computers, the 80C31 microcontroller uses the binary system internally. All variables, constants, alphanumeric characters, etc., are represented by groups of binary digits.

The 80C31 is part of a family of microcontrollers developed by Intel Corporation. The MCS-51 family became generally available around 1980. Options in the family originally included the 8051 with fixed program memory and the 8751 with an erasable program memory. The 80C31 used here requires an external memory to hold programmed instructions.

This family of microcontrollers has been commercially very successful. They are so popular that several other companies also produce them or derivatives of the original design. Unlike the case with personal computer hardware with which obsolescence occurs before the ink on your check dries, support for this chip will continue for years, I think.

The 80C31 and its relatives contain an 8-bit central processing unit. Most operations process 8-bit-wide variables. Internal random access memory (RAM) and virtually all other variables are 8 bits wide.

Unlike a general-purpose desktop computer, program memory and data memory are separate. In this case, the separation is physical in addition to having different addressing schemes. Addresses for program memory are 16 bits long, while the temporary RAM memory for data variables uses 8-bit addresses. Figure 12-7 is a block diagram of the three-chip computer.

Each port, labeled P0 through P3, of the 80C31 has eight pins which can be used to input and output information. In this case, P1 and P3 are used to communicate with the radio circuits. P0 and

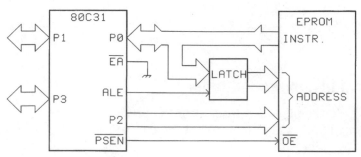

■ **12-7** *The 80C31 must get program instructions from an external memory chip. Other versions of the microcontroller have internal memory and can dispense with the other two chips.*

P2 send address values to the erasable programmable read-only memory (EPROM). P0 also receives program instructions from the EPROM.

To receive a programmed instruction from the EPROM, the 80C31 must first tell the EPROM where the address resides. Here, we need 14 bits and 14 pins to do this. The 80C31 can supply as many as 16 bits at a time for larger EPROMs.

After the desired address is accessed, the 80C31 is ready to receive an 8-bit instruction. Instead of tying up another eight pins, P0 changes roles and receives the programmed instruction from the EPROM.

In order for the EPROM to send this information, the address data supplied to the EPROM must remain present at its pins. This is the purpose of the latch. While P2 can continue to send the upper 6 bits of address data, P0 must switch jobs and become an input port. P0 can't input and output signals at the same time, so the latch holds the lower 8 bits of address information for the EPROM.

In addition to having valid data present at the EPROM's address pins, it must receive a command at the output enable pin to send instructions back to the 80C31. This command is given by the program-store-enable pin (not-PSEN) of the 80C31.

The address-latch-enable (ALE) pin of the 80C31 pulses high to trigger the latch. Timing of the ALE pulse is arranged so that it is inactive (low) before not-PSEN becomes active (low) and commands the EPROM to output a byte of instruction data.

Memory

The small, three-chip embedded computer in the transceiver uses only two types of memory devices: RAM inside the 80C31 and external EPROM. They both allow information to be accessed at random. Any address may be accessed in any order.

This is in contrast with a sequential access memory such as magnetic tape drives. To access data on such devices, portions of the tape not containing the desired data may have to be read.

As we have seen with the EPROM, applying the correct address and control signals will make data stored in the memory available. With the on-chip RAM you don't even have to worry about the routing of address, data, and control lines because they're already there.

In this system, we have a whopping 128 bytes of RAM! Granted, that does not seem like much, but it's quick and easy to access because only 1 byte is needed to address a location. RAM on the 80C31 chip is used for storing variables, which may change as the program runs.

Two main categories of solid-state RAM are commonly used in various computers. Large quantities of dynamic RAM (DRAM) are used by makers of desktop personal computers. DRAM stores a bit of information as the presence or absence of a charge. Because it is stored in a tiny capacitor, this charge leaks away. It must be refreshed thousands of times per second.

The other common type of memory, known as static RAM (SRAM), is used in the 80C31 and is better suited to our purposes. It is called *static* because memory contents remain unchanged without any outside intervention. This is true as long as power is applied.

The memory elements of static RAM can be constructed by using flip-flops. When operating in what's known as *power-down mode,* the circuits in the 80C31 can retain data in RAM. So little power is consumed that a small battery or even a large capacitor can supply enough power to maintain RAM data between operating sessions.

The 27C64 EPROM used as program memory will hold about 8 kilobytes (KB) of data. Less than half of its capacity is needed to hold the program as it stands now.

The program-memory EPROM outputs data if we feed its address and control lines what they need. How does the program data get there in the first place? Well, if you have no other recourse, you simply beg, borrow, or steal a programmed EPROM from someone who has the facilities for burning the information into the chip. Then, you plug it into your system and proceed with the project.

If you want to have more control over the embedded program, here's what happens. Field-effect transistors fabricated with a floating gate in the EPROM are used as storage elements. The floating gates can hold a charge, as would a capacitor plate.

The gate is so well insulated that a charge applied to the gate will remain there indefinitely. With no conductive path to the gate, some means must be found to apply the charge. This is done by making the insulating oxide thin enough and of uniform thickness to allow a higher-than-normal voltage to break through and charge the gate.

If the voltage is accurately controlled, this breakdown of the insulating oxide layer is nondestructive. Now that the charge is there, how is it removed? Ultraviolet (UV) light striking the gate will cause the charge to leak away, thus erasing the EPROM.

In practice, the EPROM is placed in a socket hooked to circuits specially designed to store the correct potential on each gate. This is done by addressing a group of storage elements and applying a programming voltage until the gate reaches the proper level. If you want to change something in the program, you simply pop the EPROM into a UV eraser for a couple of minutes before reprogramming it.

A variety of apparatuses capable of programming EPROMs are available. Some are capable of functioning as stand-alone units and some interface with computers. Repeatedly erasing and programming the EPROM can be rather time-consuming. If you want to experiment extensively with the program, there are easier ways to do so and they will be explained in a later chapter.

Microcontroller hardware

The following is a general but brief description of the 80C31 microcontroller. It barely scratches the surface of the world of sophistication contained in this little chip. If you wish to modify this project or design another one, you will need to acquire additional technical documentation from other sources.

Figure 12-8 is a diagram of the major functional blocks which make up the 80C31 microcontroller. It combines a central processing unit (CPU), RAM, and input/output ports.

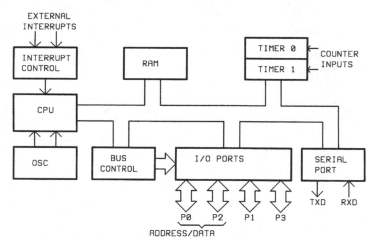

EXTERNAL INTERRUPTS

INTERRUPT CONTROL

RAM

TIMER 0

TIMER 1

COUNTER INPUTS

CPU

OSC

BUS CONTROL

I/O PORTS

SERIAL PORT

TXD   RXD

P0   P2   P1   P3

ADDRESS/DATA

■ **12-8** *Much of the 80C31 on-chip communication is handled by an 8-bit-wide data bus.*

Mode, status, and data registers are also associated with blocks listed in the diagram. An 8-bit data bus allows different parts of the 80C31 to communicate with each other. This bus is buffered to outside connections through an I/O port when memory or I/O expansion is needed.

The CPU is the brains of the microcomputer. It reads the user's program and executes instructions stored therein. The oscillator is the master timekeeper for the CPU and timers. Interrupt control (along with proper software instructions) allows the CPU to drop what it's doing when an external event occurs and start another routine in an orderly manner.

Timer 0 and timer 1 were described in chapter 10. The serial port makes duplicate use of two pins on port 3 for serial communication. With all of these functional blocks connected to each other via the internal bus, bus control must make sure that each block sends data, receives data, and releases the bus at the proper time.

Figure 12-9 is a labeled diagram of all the microcontroller pins. Labels starting with the letter P indicate port pins.

All four ports of the 80C31 are bidirectional. They can serve as output pins or input pins as commanded by software. Ports 1 and 3, which we will be using for external I/O, have three P-channel field-effect transistors (PFETs) per pin used for outputting highs as well as an N-channel field-effect transistor (NFET) that acts as a sink for outputting lows. Two PFETs are used as pull-up transistors.

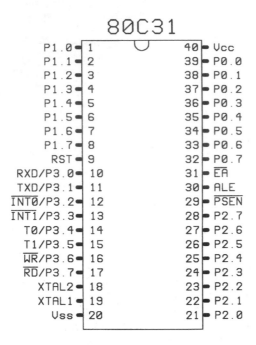

```
                    80C31
         P1.0  1          40   Vcc
         P1.1  2          39   P0.0
         P1.2  3          38   P0.1
         P1.3  4          37   P0.2
         P1.4  5          36   P0.3
         P1.5  6          35   P0.4
         P1.6  7          34   P0.5
         P1.7  8          33   P0.6
          RST  9          32   P0.7
     RXD/P3.0  10         31   EA
     TXD/P3.1  11         30   ALE
    INT0/P3.2  12         29   PSEN
    INT1/P3.3  13         28   P2.7
      T0/P3.4  14         27   P2.6
      T1/P3.5  15         26   P2.5
      WR/P3.6  16         25   P2.4
      RD/P3.7  17         24   P2.3
        XTAL2  18         23   P2.2
        XTAL1  19         22   P2.1
          Vss  20         21   P2.0
```

■ **12-9** *This pin-out of the microcontroller is similar to a view of the top side with the pins facing down-ward and away from the viewer.*

One pull-up transistor emits a strong current pulse to speed up low-to-high transitions at the pin. The other pull-up transistor emits a weaker current continuously as long as a 1 is written and latched to the port pin.

When a software instruction reads the status of a port, some software commands read the actual level present at the port pin, while other instructions read the status of an internal latch connected to that pin.

If you decide to drive a transistor in a common emitter configuration by connecting the base directly to a port pin, the actual pin voltage would always appear low. Reading the internal latch allows the microcontroller to determine whether the transistor is being turned on or off.

All of the port 3 pins are multifunctional. We have already seen in figure 10-8 how port 3, pin 2 is used to control a timer/counter circuit and generate a hardware interrupt. Port 3, pin 3 has similar capabilities. Only one of the timer/counter sections is used in the transceiver controller (so far). P3.3 is used to monitor the key line to tell if the radio is in transmit or receive mode.

RXD at port 3, pin 0 and TXD at port 3, pin 1 are part of a versatile serial interface. If the peripheral devices communicated with are compatible, this interface can provide a fast and easy method of data transfer. Unfortunately, register size on some of the transceiver chips is not compatible. A slower but acceptable software routine generates data and clock signals for these two pins.

Port pins 3.4 through 3.7 are used as regular input/output pins and do not use any of the alternative functions. These unused functions are T0, T1, not-WR, and not-RD. T0 and T1 are external input pins for timers 1 and 0. Not-RD is for driving the output-enable pin when reading an external RAM chip. Not-WR would go to the write-enable pin of an external RAM chip.

XTAL1 and XTAL2 can be connected to a quartz crystal and two capacitors to operate the on-chip oscillator. Alternatively, a signal from an external oscillator can be connected to XTAL1 as it is used in the transceiver.

The computer board

To give the radio its "intelligence," we only need to build a three-chip computer with a few accessory circuits. In figure 12-10, the actual computing hardware consists of IC1, IC2, and IC4.

IC1 is, of course, the microcontroller. IC2 is an 8-bit latch. IC4 is an EPROM which contains the programmed instructions.

Because of the need to send a portion of the CPU clock signal to the synthesizer mixer, an external oscillator built of discrete components is used at Q1. This is the circuit as it exists in the prototype.

Crystal Y1 is a mass-produced unit designed to be used at 9.8304 MHz. This circuit is actually operating at 9.831396 MHz. Using the external circuit makes it easier to operate the crystal approximately 1 kHz above its design frequency.

Unfortunately, doing so compromises frequency stability because of temperature-induced drift. This can be corrected to some extent by experimenting with the temperature coefficients of C3 through C6.

A better solution is to order a high-quality crystal ground for the correct frequency. You should probably use zero-temperature-coefficient capacitors at C3 through C5. An air-insulated ceramic trimmer is a good choice for C6. Try to keep the connection from C2 to the mixer as short as possible.

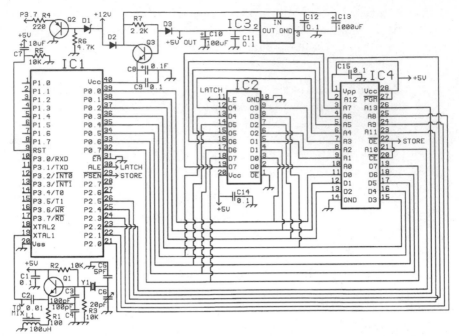

■ **12-10** *This is the schematic of the three-chip computer which makes up most of the transceiver's controller board. C8 is a 100,000-µF memory backup capacitor such as Mouser stock number 551-FR100 or similar. See text for alternatives. D1 is a 6.2-V zener. D2 is a 1N4001 or similar. D3 is a 1N4148 or similar. IC1 is an 80C31. IC2 is a 74LS373. IC3 is a 7805 three-terminal regulator. IC4 is a 27C64 EPROM. JDR Microdevices and Jameco are possible sources for microcontrollers and ICs. Digi-Key and Circuit Specialist also carry some of the ICs. Suggested transistors are: for Q1, a 2N2222A; for Q2, a 2N2907 or similar; for Q3, use an SK3444 or other transistor with a high emitter to base breakdown voltage.*

The other circuitry on the board relates to power issues. The computer must receive clean, well-regulated power but it has some additional requirements. The main 12-V supply is routed through D2 and IC3 to supply 5-V power for everything on the board but IC1.

When power is first applied, the 80C31 needs to put its own internal affairs in order by going through a reset procedure. If it doesn't do this, some internal registers will have incorrect data and the program may crash.

Components C7 and R5 take care of this potential problem by applying a momentary pulse to the reset input at pin 9. After C7 is fully charged, pin 9 stays at ground potential during operation.

When 12-V power is removed, powering down the radio, a couple of other chores need to be undertaken. The microcontroller must be notified that main power is gone so that it can convert to power-down mode. When this is achieved, it will retain RAM information while drawing less than 10 μA.

As long as 12-V power is applied, current flowing through D1 biases Q2 off. If the 12-V supply goes away, the computer can function for a short while because the charge in C13 supplies regulator IC3.

When Q2 turns on because of a loss of 12 V, pin 7 on port 3 is pulled low. A software routine periodically polls P3.7 to check its status. If it finds it has been pulled low by Q2, the power-down sequence is initiated and the 80C31 effectively "goes to sleep."

The prototype transceiver uses C8 to power the on-chip RAM in this condition. This small capacitor provides enough power to retain RAM data for about 8 to 10 days. Q3 provides regulated power to pin 40 of the 80C31. It follows the regulated output voltage of IC3.

When the 12- and 5-V supplies are eliminated, the emitter junction of Q3 is reverse-biased. Almost all of the charge in C8 is routed to pin 40. The voltage at pin 40 can fall to 2 V or even less before RAM data is lost.

If you wait too long between operating sessions and the information in RAM is lost, only frequency and mode data are lost. These are easily entered back into memory from the front panel. All operating system instructions are stored in the EPROM.

You can easily extend the time period for supplying backup power to IC1. One obvious way is to use larger values for C8. This could be accomplished by mounting larger capacitors or even banks of them off-board.

Uninterrupted memory power could be supplied for years by a small battery. The typical choice for an application like this would be a small, sealed, inorganic lithium battery.

Non-rechargeable batteries cannot be connected directly in place of C8. They must have a protective diode in series to prevent current from flowing into the battery.

Vanadium pentoxide rechargeable lithium batteries are now available and will need a series resistor to limit charging current. Consult the manufacturer's specification for charging rates.

If you decide to use any other kind of memory backup power, I recommend placing the battery outside the radio compartment. The best place would probably be on the bottom of the enclosure. Some kinds of cells may leak, and I'm sure you don't want the corrosive contents dribbling on the insides of your radio.

A physical layout of the transceiver controller board is shown in figure 12-11. With this board, components aren't terribly numerous, so stuffing parts in the wrong place shouldn't be as much of a hazard.

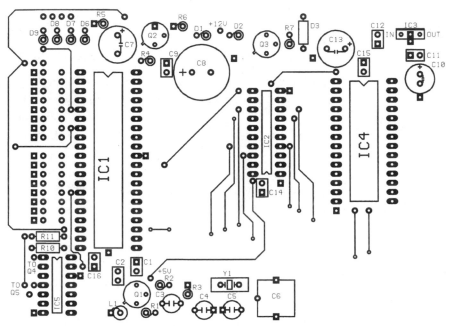

■ **12-11** *This is a component-side layout of the controller board. It is a source of electrical noise and should be isolated from the radio circuits.*

To eliminate the need for etched paths on the top side, a number of jumper wires are used. Be especially careful to route these correctly. Don't forget to countersink holes as necessary.

Various header connectors are available which will fit the pads connected to ports 1 and 3. Some of them have a double-row pattern which will allow every other wire in a ribbon cable to carry a ground connection from the row of square pads.

This type of cable provides a low-impedance ground plane between the boards and helps reduce radiated noise from the signal lines. If you use such a connector and cable, drill extra holes and jumper each line of square pads to the ground plane.

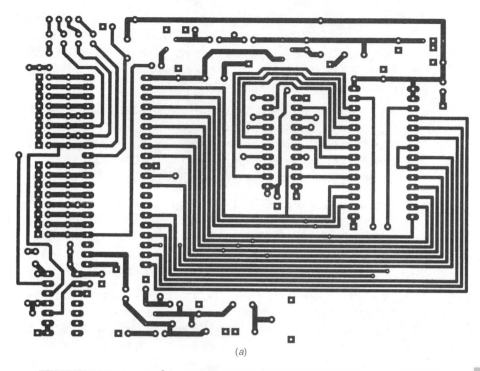

(a)

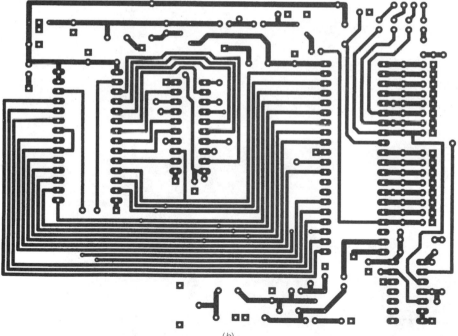

(b)

■ **12-12** *(a) A mirror image of the etching pattern, and (b) a direct view of the pattern (actual size).*

Although not described in this chapter, IC5 is used for interfacing with a system for monitoring the rotation of the main tuning knob. R11 and R10 are part of this system. Q4 and Q5 phototransistors are not mounted on the board because they pick up light pulses from an encoder disk on the shaft knob.

Etching patterns are displayed in figure 12-12. The view at *a* is looking through the ground plane; *b* is a direct view of the paths. Spacing between pads and lines where some of the jumper wires connect is very close. Be sure to use a small enough drill bit here: a diameter of 0.018 in should be suitable.

Use insulated wire AWG number-26 or smaller for the jumper wires. It's probably best to install the jumpers ahead of all the other components. After you have them soldered in place and visually inspected, a quick check with an ohmmeter can be used to find any hard-to-spot bridged connections or shorts.

Of course, you should use sockets for the ICs. If you have an EPROM programmer and plan on trying some changes, you might even want to use a zero insertion force (ZIF) socket at IC4. ZIF sockets are expensive, but they allow you to insert and remove the IC with little effort. They usually have a little lever or handle which you use to tighten or loosen connections to the pins.

Operator-to-transceiver interface

13

MANY TRANSCEIVER FUNCTIONS ARE EFFECTED BY DATA flowing to and from the controller board, as described in the preceding chapter. Another, even more important, path for information and control is that which exists between the operator and transceiver.

To successfully operate the transceiver, we need to exchange information in two directions. Not only must the operator be able to affect circuits in the radio, but information about the internal status of various transceiver workings must be available to the operator.

A liquid-crystal display (LCD) provides the operator with information about various operating parameters such as frequency and mode. Devices on the front panel used by the operator for inputting information include push-button switches and a rotary shaft encoder.

In chapter 12, you may have noticed that some pins on port 3 had connections to multiple ICs. This is possible because each IC is communicating with the port pin at different times. They are time-multiplexed.

Instead of using only high and low logic states, a third high-impedance state is used. For example, three hypothetical devices shown in figure 13-1 share a common 4-bit bus. In this case, they are all output devices, but the same technique can be used for inputs.

Each set of four switches would be controlled by an external signal, probably one emanating from other microcontroller port pins. As long as the three microcontroller pins controlling switches in devices A, B, and C are timed separately, each device can send data along the bus without conflicting with the other two.

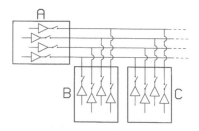

■ **13-1** *An example of three devices sharing the same four-wire bus by using it at different times.*

A conflict arises when a driver in one device outputs a logic level, while a driver in a different device tries to output an opposite level on the same line. Solid-state switches are not always required for the high-impedance state. Although many ICs are available with such tristate capabilities, you will later see that diodes can sometimes be used to accomplish the same end.

Optical shaft encoder

Tuning the transceiver to various operating frequencies is accomplished by turning a knob mounted on an optical shaft encoder. Doing so indirectly changes the value stored in a 3-byte memory location. This binary number controls the synthesizer. It is also converted eventually to decimal digits on the LCD.

This optical shaft encoder uses a small rotating disk with cutout segments which pass or block infrared light beams as the disk rotates. Patterns shown in figure 13-2 can be used to make your own disk. The disk shown in *a* is similar to the one in the prototype transceiver. The disk in *b* requires less fabrication and has half the tuning rate.

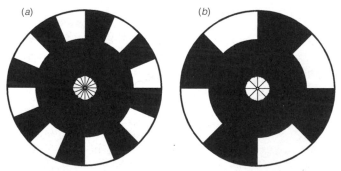

(a)   (b)

■ **13-2** *You can make a paper copy of the disk and paste it on a piece of sheet metal to use as a cutting template (actual size).*

Disks can be made from a variety of materials; however, unsuitable ones may not be so obvious. Infrared light passes through many kinds of plastics. Materials opaque to visible light may not always block infrared light sources. One solution to this problem is to use metal for the disk. It doesn't matter what kind; just use whatever is convenient and easy for you to cut. I used some thin aluminum.

You may even be able to take advantage of the infrared properties of some plastics. You might want to try etching the figure 13-2 patterns on printed circuit board material. Some board substrates such as Mylar are rather clear. This would eliminate the need to cut slots or teeth along the periphery.

If you cut the disk from metal, a hand shear (snips) can be used for most of the work. Cutting the inside edge of the slots requires some other method. A nibbling tool could be used or you could drill a line of small holes and break the piece loose by bending it.

After drilling a ¼-in-diameter shaft-hole, the toothed disk is painted with flat-black enamel. It is mounted on a ¼-in-diameter bolt. Select a bolt with an unthreaded portion long enough to pass through the front panel bearing and into a knob.

To prepare the bolt, cut off the head and remove rough edges with a file. I also center-drilled the threaded end and tapped the hole for a small machine screw. This allowed me to mount a washer which bears against the back bearing. The washer limits how far forward the shaft can slide. The knob and another washer limit backward movement.

An easier method of making the stop might be to use two more jam nuts on the threaded portion of the bolt. If there's insufficient room for this method and you need to drill a hole in the end of the bolt, here are some tips.

The best method is to mount the bolt in the chuck of a metal-cutting lathe. First, use a small cutting tool to cut a smooth face on the bolt end. By positioning the cutting tool correctly, you can make a cut which bevels inward to a point accurately located at the center of rotation. Drill the hole by mounting a drill bit sized for the tap in the tailstock.

Many builders do not have access to a lathe. You can also do an acceptable job with a drill press or even a powered hand drill. The hand drill must be clamped securely in a stand. Locate and indent the bolt's center of rotation as accurately as possible with a marking punch.

Chuck the bolt into a drill press or table-mounted hand drill. With a drill press, you can mount the drill bit in a vise supported by the drill table. Make sure the bit is in line with the bolt and not tilted.

To use a hand drill, clamp the drill bit in a pair of locking pliers and carefully feed it in by hand. This is the most difficult method and requires patience. Regardless of the method, remember to wear proper eye protection and follow all applicable safety precautions.

Mount the disk by threading on a matching nut. Slide the disk on the shaft and follow it with another nut, as shown in figure 13-3. Fasten the disk by tightening both nuts against each other.

Supporting the front end of the rotary shaft is a modified ¼-in phone jack. Cut off the tip contact, the part that normally presses against the phone plug end. The rear shaft support in the photo is a spherical rod-end bearing.

Spherical rod-end bearings are rather expensive for such a light-duty application. Almost any type of rear bearing you devise will be suitable if you take care to align it and keep the shaft from binding. The spherical bearing is self-aligning, within limits, and is therefore a bit easier to use.

■ **13-3** *This is the optical shaft encoder assembly. The bundle of wires which fan out at the left connect to LEDs and phototransistors.*

The infrared LEDs and photodetectors are mounted in a plastic housing with and aimed directly at each other. The housing has a slot through which objects such as the disk teeth can move. As the teeth pass through the slot, they interrupt the light beam, which drastically reduces current flow in the phototransistors.

Another possibility is to use visible light. You could purchase or devise an interruptible photocoupled pair from red LEDs and phototransistors. One reason for the widespread availability of infrared photocoupled pairs is that they tend to be more robust when used in suboptimal conditions.

Infrared LEDs are such efficient light emitters that the beam is less easily obscured by dust. Other effects, such as lower output caused by aging, are less problematic.

The photocoupled pairs must be supported and each pair has two mounting holes for this purpose. You can fabricate supports in various ways. I chose to make brackets from AWG number-12 bare copper wire. The wire is bent into a shape necessary for attaching the photocoupled pair and soldered to lugs on the ¼-in phone jack.

Rotation of the tuning knob is measured by counting pulses from the phototransistors. That's fine for measuring how much rotation has occurred, but it gives no information about the direction of rotation. Fortunately, direction can be determined by using two interrupted light beams at different locations along a circular path.

To see where these light beams need to be placed, refer to figure 13-2. Let's call the dark areas protruding from the hub *teeth* and the light areas in between them *spaces*.

It's probably best if the spacing between light paths is described as an angular measurement. Each tooth in disk B is 45° wide, while disk A has teeth covering a 22.5° angle. Let's name the width of a tooth T in either case.

A simple rule for making this system work is to use a spacing of 1.5T, or 2.5T, or 3.5T, etc. To measure the correct spacing, one light path should line up with the edge of a tooth. This also means it's bordering on a space. A spacing of 1.5T is the same as spanning across one tooth and halfway into the next space.

For example, the closest possible spacing for light paths should be a separation of 67.5° on disk B. This would also be 1.5T. A spacing of 2.5T on disk A equals 56.25°.

Actual spacings don't have to be accurate within fractional portions of a degree or even a few degrees. More error will cause the step points to be more unevenly spaced, but the system will still work.

You can mark the correct spacings on a piece of paper and drag it around the corresponding disk pattern. Assign a logical low when a mark lands on a space. Assign a high level to the mark when it is on a tooth.

If you carefully note the pattern generated by doing this, it will match the binary pattern labeled "activation sequence" at the top of figure 13-4. This sequence also contains the first four numbers in the gray code, a binary code in which only one bit changes at a time.

An important fact is that the order of binary values for clockwise rotation is different from those generated by counterclockwise rotation. By knowing the current and previous value, you can determine the direction of rotation.

As an example, if the current value is 11 and the previous value was 01, the shaft rotated one step clockwise according to the activation sequence at the top of figure 13-4. If the current value is 00, it rotated one step counterclockwise. If the indicated sense of direction is opposite to what you desire, simply switch the photodetector wires to reverse it.

The diagram in figure 13-4 shows how each case is tested by a software routine to determine how the current and previous values compare. *M* represents a previous binary value stored in memory. *P* represents the current binary value at port 1.6 and 1.7 (pins 7 and 8 of the 80C31).

The first case to be tested asks if the previous value stored in memory was 01. If not, another value, 11, is tested against memory. If the preceding value in memory was indeed 01, the current port value is tested.

A value of 00 causes the routine to decrease the operating frequency, while a value of 11 causes it to increase. If neither value is present, the current value must be out of sequence and no action is taken. In any case, the present value is stored in memory and the microcontroller goes on to run other routines until it again revisits this one to repeat the same process.

Other values in figure 13-4 are tested in a similar manner. Other routines determine the amount of increase or decrease in frequency per step and can be selected at the front panel. Although it's not shown in the diagram, a necessary part of the routine is to

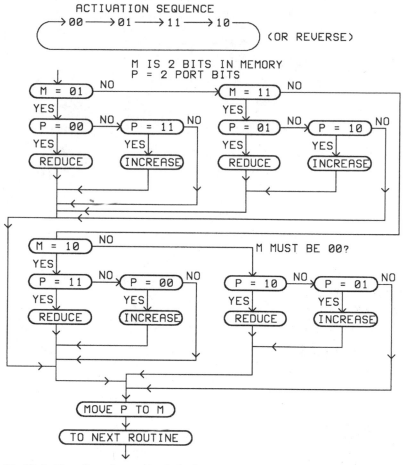

ACTIVATION SEQUENCE

→ 00 ——→ 01 ——→ 11 ——→ 10 ——

(OR REVERSE)

M IS 2 BITS IN MEMORY
P = 2 PORT BITS

■ 13-4 *Reading the optical shaft encoder is just one of many tasks performed by firmware running on the 80C31 three-chip computer. Charting information flow was the first step in writing this particular routine.*

switch the phototransistors in and out to prevent contention with other devices on the bus.

Along with other circuits, a schematic diagram of the optical pulse-counting components is shown in figure 13-5. Q4 and Q5 are the phototransistors. Each uses a pull-up resistor connected to the collector. Diodes D4 and D5 are infrared light-emitting diodes connected in series with R9. Half of a 4066 is used at IC5 to multiplex the phototransistors onto the bus.

P3.4 controls switches A and B in IC5. These electronic switches bring the phototransistors on line when P3.4 is high and takes them off when low.

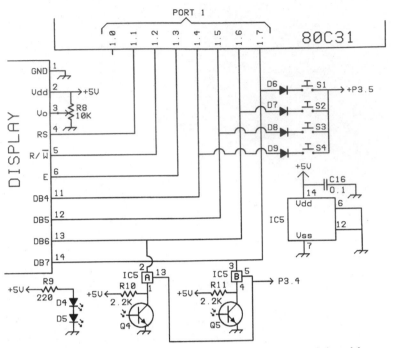

■ 13-5 *D4 and D5 are LEDs in optical interrupter modules, Mouser stock number 512-H21B1. D6 through D9 are 1N4148 silicon diodes. IC5 is a 4066. Q4 and Q5 are part of the interrupter modules. S1 through S4 can be any kind of normally open, momentary-contact, push-button switch.*

Pull-up resistors R10 and R11 serve two purposes. They decrease the rise time of collector voltage. They also cause the phototransistors to be less susceptible to activation by ambient light. It is certainly easier to adjust or test your transceiver if it's not being blinded by the work light when the cover is off.

Some builders may acquire optically coupled interrupter modules without information about the lead arrangement. You can determine the phototransistor end by using an ohmmeter and light source. Even though the phototransistor may be partially blocked by the matching LED, a 60-W incandescent bulb will provide sufficient infrared light for the test.

If you are using a very high impedance meter, such as a modern digital multimeter, it may not supply enough current to be sure of the correct polarity once you've identified the phototransistor. If so, use it to measure current through a 2.2-kΩ resistor connected to a 5-V supply.

Once you have the phototransistor operational, use a 220- or 270-Ω resistor and 5-V supply to power the LED. When you have the correct polarity, it will be indicated by a large increase in current flow through the phototransistor.

Panel switches

Another important operator interface uses front panel push-button switches. Four momentary, normally open switches are labeled S1 through S4 in figure 13-5. These switches are time-multiplexed with the previously mentioned phototransistors and an LCD module.

Diodes D6 through D9 isolate each port pin from the other three pins. When P3.5 is low, the highest bits of port 1 can read the status of the switches. At other times, this 4-bit bus may be outputting various levels to the LCD module. If not for the diodes, multiple switch closures could short bus lines together and corrupt data.

Part of the software routine requires that button presses be valid for a fixed length of time or the switch closure will not be recognized. The validation process is necessary for several reasons. The contacts of mechanical switches often open and close rapidly before reaching a steady state. This rapid burst of indecision happens when the switch is pushed or released. It is usually referred to as *contact bounce*.

Two other reasons for the validation period relate to how the buttons are used. First, some functions require pressing two or more buttons at once. It is extremely difficult to consistently manually close two switches at exactly the same time. Running at full speed, the microprocessor could easily interpret this action as two separate incidents.

Some buttons can be used in a second way, which requires a delay. Memory-up, memory-down, and tune rate can be activated by a single press or by holding the button down continuously. Pressing one of these buttons continuously causes the software to page through all of the available choices or tuning rates. If the paging process jumps from one selection to another too rapidly, it is impossible to follow the choices.

The push-button switches are used for a number of functions. Figure 13-6 shows the physical arrangement of these switches on the prototype transceiver's front panel. Except for numbering, it is also labeled in a similar manner.

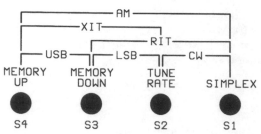

■ **13-6** *You will probably want to label the push-button switches on your transceiver. It's a lot easier to show it off to guest operators when they can quickly find out how to change mode or operating frequency.*

Labels closest to the switch buttons indicate that the operator enters information by pressing the button directly below. Those functions which require only a single button are memory-up, memory-down, tune rate, and simplex. The second line of labels includes upper sideband, lower sideband, and CW (continuous-wave telegraphy). Selecting one of these emission modes requires a two-button activation.

XIT and RIT stand for *transmit incremental tuning* and *receive incremental tuning.* Split tuning might be a better moniker for these functions. Activating XIT fixes the receive frequency at the current setting and makes the transmit frequency tunable. RIT fixes the transmit frequency at the currently displayed value and makes the receive frequency tunable.

To restore normal simplex operation in which transmission and reception occur on the same frequency, simply press S1. If you build AM capability into your transceiver, it is selected by simultaneously pressing the two outside push buttons, S4 and S1.

Memory-up and -down select locations in a memory bank which contains frequency values entered by the operator. Memory locations are named alphabetically on the display. The current program uses 16 locations named A through P.

Each location may be programmed to any frequency within the transceiver's range of operation. When you switch from one frequency to a second one by pressing the memory-up or -down buttons, the last operating frequency at the first memory location is remembered until you return to the first memory location.

As an example, suppose you are tuning around on 75 m, using memory location B. After listening to an SSB conversation at 3900.31

kHz, you punch the memory-up button until memory F is active. Location F holds a frequency value of, say, 14,035.65 kHz where you made your last CW QSO. Next, you tune down to 14,029.00 kHz.

If you return to location B, the transceiver will again be tuned to 3900.31 kHz. Returning to location F, the transceiver is now tuned to 14,029.00.

Tuning rates for optical shaft encoder inputs are selected by S2. Presently, four rates are available. You can select increments of 10, 100, 10,000, or 1,000,000 Hz per step by pressing S2. In corresponding order, these increments are displayed as s, m, f, and h as abbreviations for small, medium, fast, and huge.

Perhaps the most unusual push-button function is the one that puts the display into a diagnostic mode. To do this, you must push all four buttons at once. This is an awkward maneuver and is intentionally designed to be so. This is not the normal operating mode of the transceiver. It is used for troubleshooting or adjusting circuits in the frequency synthesizer.

The diagnostic display mode allows you to see the synthesizer's mixer output frequency, target frequency, and the DAC input value. To take the display out of diagnostic mode, you must punch the simplex button and power-down the transceiver. More details about the diagnostic display mode will be given in the next section, where we examine the LCD module.

LCD and controller module

A major portion of information concerning the transceiver's status is available to the operator through a liquid-crystal display module. The photo of the display in figure 13-7 shows how it looks installed and operating on the prototype.

A capital "I" at the upper left of the display indicates which location in the frequency memory bank is active. Directly below, at lower left, the letters "USB" indicate that the transceiver is in upper-sideband mode.

Letters "TX" and "RX" are in line with their respective transmit and receive frequencies. In this case, both frequencies are the same. When using split frequencies, no operator action is needed to see the inactive frequency, because both are displayed simultaneously.

In simplex mode, using the same transmitting and receiving frequencies, the "TX" and "RX" letters disappear. Both frequency displays remain. They are duplicates of each other.

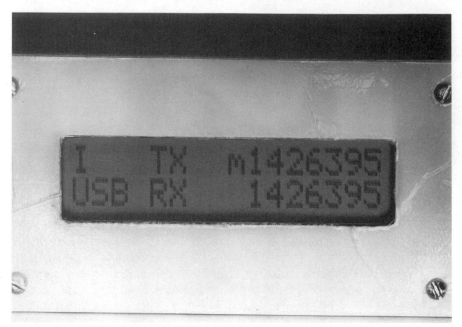

■ 13-7 *This is how the display looks in operation. It has the capacity to display two rows of 16 characters each.*

The lowercase "m" to the left of numerals 1426395 indicates that the medium or 100-Hz-per-step tuning rate is in use. The frequency displayed in this example is 14,263.95 kHz in the 20-m amateur band.

When the diagnostic display mode is activated, some parts of the display are replaced by new figures and some remain undisturbed. Undisturbed information includes emission mode, memory location, tune rate, and the upper-right frequency display area.

"TX" in this example would be replaced with a group of decimal digits indicating the value sent to the digital-to-analog converter. "RX" is replaced by digits reporting the synthesizer mixer frequency as measured and calculated by the computer.

The lower-right space displaying the receiving frequency would display the target frequency for the synthesizer's mixer output. This is the frequency which the fine-tune loop tries to achieve in the output of its mixer.

When you are through with the diagnostic mode and want to return to normal operation, two steps are necessary. Push the simplex button and disconnect power to the transceiver. Normal display operation will be restored when you power up the transceiver.

A major reason for using LCDs instead of some other technology is the issue of power consumption. In the past, radios have used a number of display technologies including LEDs and vacuum fluorescent displays. To display as much information as the LCD module would require enough power to become a significant portion of the total draw while receiving.

LCDs are easily viewed under high ambient lighting conditions and require little power. They also need to be driven by an ac source. If the segments or elements are driven by direct current, the display will eventually be irreparably damaged.

The ac waveform is often a symmetrical square wave because LCDs are commonly driven by logic circuits of one sort or another. Figure 13-8 shows an example of how an exclusive-OR gate can be used to drive a single element of an LCD.

An exclusive-OR gate is like a plain OR gate except that when both inputs are true, the output is false. Here, it is useful because when one input is constant, the output is always a negated version of the other input.

A controlling input can cause the output signal to be in phase or out of phase with the square-wave input. When the LCD element and LCD common connection are driven in phase, no voltage difference exists and nothing appears at the element location. When driven 180° out of phase, a square waveform across the element results, making the element visible.

Fortunately, we don't have to build the driver circuits because they and a lot of other functions are built into the module. The

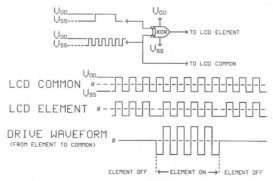

■ **13-8** *The exclusive-OR circuit is often used to drive LCDs with a square-wave ac signal. The square-wave signal connected to LCD common is usually a 30- to 100-Hz signal generated by a constantly running oscillator.*

LCD module is a dot-matrix type with its own microcontroller which communicates with the transceiver's computer.

The display in the prototype is from Mouser Electronics, stock number 592-RCM7037X. Many similar displays compatible with this project are available from other suppliers. I highly recommend using a backlit version of the display. They are really nice looking. Unfortunately, they are also slightly thicker and I didn't allow sufficient room for installing one.

Data can be sent to the display as two cycles of 4-bit data or one cycle of 8-bit data. In this application we will use the 4-bit method. The display is commanded to use the 4-bit mode by a software routine when the transceiver is first turned on.

Notice that pins 7 through 10 are not listed in figure 13-5 because they are unused. Four-bit data transfer between the display module and 80C31 is completed when the four higher-order bits are transferred first, followed by the four lower-order bits.

The timing diagram in figure 13-9 shows that, basically, the 80C31 software checks to see if the module is busy, tells the module where to put a character, checks again for a busy condition, and tells the module which character to display.

The waveforms are those generated by the 80C31. The module-generated busy flag comes from the DB7 line and will appear as a high if the module is not ready to accept data. If the 80C31 read

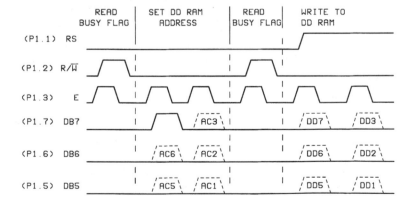

■ **13-9** *The waveforms here are not to be taken literally. They illustrate the relative timing used when communicating with the display module.*

routine sees this line is high, it waits until DB7 goes low before sending data to the module.

The command to set the display data (DD) RAM address is clocked in by two cycles of the enable line. The RAM referred to here is inside the display module. This 6-bit value, which specifies the location of a character, sets the value of an address counter in the display module.

A command in the initialization routine at power-up has already told the module to decrement its address counter with each new character. Characters in this case are being written right to left.

For this reason, a set-DD-RAM address command is not always necessary with each new character because some character data is sent as a continuous group. Often, the new character will not be in line with a previous group. The 80C31 sends a new DD-RAM address when a new group with a different starting location is needed.

"Write to DD RAM" is a command sent from the 80C31 that specifies which character is to be displayed. The module contains a variety of character patterns stored in ROM (read-only memory). The 8-bit number representing each character in this application can be found in an ASCII (American Standard Code for Information Interchange) table.

For a more concrete example of bit values when a character is displayed, see figure 13-10. Like the initialization procedure, turning the display on is done at power-up. One difference between this example and the transceiver's display is that this example turns on the cursor to show where the character will be located.

The cursor is turned on or off by databit 2 (DB2) when the display-on/off command is sent. Let's assume for the moment that the display module is not busy. Lines connected to DB0 through DB6 specify the display-data address. Being zero, the character will be located at the top left position.

When the 80C31 sends the binary number 01000001 to DB0 through DB7, an uppercase "A" appears on the display. The cursor is shifted one position to the right. In this case, the address counter was previously instructed to increment, opposite the instruction used in the transceiver software.

Figure 13-11 shows DD-RAM addresses for the remaining character positions. The display module has more capabilities than those described here. Suppliers that carry the modules often have data sheets or books with comprehensive descriptions.

COMMAND						DISPLAY	OPERATION
INITIALIZATION							GETS INSTRUCTIONS FROM 80C31.
DISPLAY ON/OFF CONTROL							TURN ON DISPLAY AND CURSOR.
RS	R/W	DB7	DB6	DB5	DB4		
0	0	0	0	0	0		
0	0	0	1	1	0		
SET DD RAM ADDRESS						—	SET RAM ADDRESS TO POSITION CURSOR AT HEAD OF FIRST LINE.
0	0	1	0	0	0		
0	0	0	0	0	0		
WRITE DATA TO CG/DD RAM						A_	WRITE "A_". CURSOR INCREMENTED BY ONE. SHIFTS TO RIGHT.
1	0	0	1	0	0		
1	0	0	0	0	1		

■ **13-10** *ASCII code 01000001 (binary) displays "A."*

CHARACTER POSITION	1	2	3	4	5	6	7	8	9	10	11	12	13	14	15	16
DD RAM (HEX) ADDRESS	00	01	02	03	04	05	06	07	08	09	0A	0B	0C	0D	0E	0F
CHARACTER POSITION	17	18	19	20	21	22	23	24	25	26	27	28	29	30	31	32
DD RAM (HEX) ADDRESS	40	41	42	43	44	45	46	47	48	49	4A	4B	4C	4D	4E	4F

■ **13-11** *Each location on the display has its own address.*

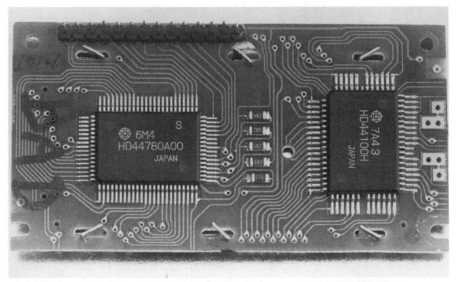

■ **13-12** *Here is a view of a module which was replaced. Use care in mounting the board because some circuit paths come relatively close to the mounting holes.*

Figure 13-12 is a photo of the back of a typical module. Some use quarter-size outline packaged IC chips. On others, the chip carrier is mounted directly on the board and covered by a drop of epoxy.

Adding a 0.1-in spacing pin header is helpful when connecting or changing modules. Pin 1 is near the end of the board. Pin 14 is near the middle.

For contrast control, a small plastic-bodied trim pot can be glued to the connector. If you don't see anything on the display, try adjusting R8 from figure 13-5. If contrast is set too low, no characters will be visible.

This can be helpful when you are not sure whether a problem lies in the display or the three-chip computer. If the display is bad, usually no change will occur as R8 is adjusted. A good display hooked to a bad computer will at least darken all of the character blocks when contrast is set to maximum.

Programming

14

IF YOU BUILD THE TRANSCEIVER AND USE A PREPRO-
grammed EPROM, no programming knowledge is required. There
are certainly plenty of hardware issues to occupy the attentions of
a builder who wishes to experiment with, modify, or improve the
transceiver.

Many details of how the transceiver operates can be changed by
someone who wants to make software changes. This is probably
true to a greater extent than with some radios, because the microprocessor is directly involved with the synthesizer fine-tune loop.

Hardware changes may also require software modifications to
accommodate new or modified circuits. Adding 6-m coverage, as an
example, would be an ambitious undertaking. However, it should
be feasible because of the 72-MHz IF and broadband circuitry.

What is a computer program?

Computers accomplish their task with such speed and accuracy
that it often appears that they are doing many things at the same
time. This is mostly an illusion. Although some processes may be
accomplished simultaneously, the illusion occurs because so many
actions are being taken faster than humans can observe.

Someone with no knowledge of computer programming playing a
video game may not realize this. When their intergalactic cruiser is
attacked near planet Zork, it looks as if all 49 Zorkonian ships are
moving simultaneously. The independent movement of 49 separate objects on the screen is really the result of many calculations
performed very quickly, one after another.

If each set of calculations is completed between screen updates,
an apparently smooth and continuous motion is perceived by the
viewer. Persistence of vision helps make the illusion seem real.
The transceiver's computer may also appear to control several
tasks simultaneously, but they are actually handled sequentially.

A computer program consists of an ordered sequence of specific steps to be executed one at a time. This collection of steps, or method used as a solution to an application, is called an *algorithm*.

The program is stored inside the computer's memory as a sequence of binary numbers. Each number corresponds to one of the basic operations which the CPU is capable of executing. In the 80C31, each number is 1 byte (8 bits) wide.

The sequence of binary numbers understood by the computer is called *machine language*. People who write programs almost never deal with machine language. Even though you can use tables to look up the number for various actions, it is a very tedious and error-prone way to write a program.

An important capability used by most computer programs is *conditional branching*. Tests can be performed on data which allow decisions to be made about the order in which instructions are executed. Branching, or jumping to a new set of instructions, makes machine language even more complicated. The new destinations must be calculated, affecting some instructions.

Another complication occurs because of addressing modes. An address in memory can often be specified in various ways. It may be indicated immediately by a number in a register.

In a different instance, it may be specified indirectly. A different register or memory location could contain a number specifying an address to be used by the CPU.

With many different instructions using a number of different modes, the possible combinations mushroom to a size impractical for manual programming using machine language.

To make the task of writing a computer program more manageable, various programming languages have been devised which perform much of the detailed work necessary to instruct the machine.

The next step up from manually managing binary computer instructions is to use an *assembler program*. Assembly language is constructed around the use of words and concepts rather than the use of encoded numbers.

Groups of numerical instructions which perform the same type of function are given a name. These mnemonics, or memory aids, usually look like abbreviations of the instructions they represent.

Assemblers for the 80C31 (and 8051 family) have the mnemonic "ADD." Yes, it simply means "add two numbers together." One number must be in the accumulator and the other is a byte variable. Although four different addressing modes are possible with the ADD instruction, a programmer needs to remember only one mnemonic.

Another mnemonic example is MOV. Its function is to move a byte variable from one location to another. Fifteen combinations of source and addressing modes are allowed. In these and other instructions, some register addressing modes have additional data embedded in the instruction, determined by which register is being used.

The main point is that an assembler can allow programmers to write a series of instructions more quickly with less chance of errors. Assemblers can also assign memory address values to labels.

Instead of having to know the address of a destination in the sequence of instructions, you could simply label the destination "THERE." The assembler will determine where THERE is so that you don't have to count all the intervening addresses used by the program between the branch point and THERE.

Assemblers also automate other chores, such as number-base conversions and defining constants. Assembly language does many things for the programmer; however, higher-level programming languages have been developed which are oriented toward the user or a specific task.

High-level languages use instructions which are constructed from a number of primitive machine-language instructions. In some cases, one line of high-level code may replace hundreds of lines of assembly code.

BASIC is an example of a high-level programming language. It stands for "Beginner's All-purpose Symbolic Instruction Code." If you've never tried any kind of programming, BASIC is a good place to start. Most IBM-compatible personal computers have some form of BASIC included with their operating system files.

Many BASIC keywords have a meaning in English very similar to the functions performed in a program. You can have instructions to READ some DATA. You might instruct the computer to DO this WHILE a certain condition exists.

A simple radio application for calculating values to the nearest kilohertz might look something like the following:

```
10 LET I.FREQ = 9000
20 LET R.FREQ = 14000
30 OSC = R.FREQ - I.FREQ
40 PRINT OSC
```

The numbers at the left are line numbers. Line 10 sets the variable I.FREQ (for intermediate frequency) equal to 9000. Line 20 sets the variable R.FREQ equal to 14,000. Line 30 computes the difference, storing it in the variable OSC. Here, it's printed on the video display. It could be used to control a frequency synthesizer.

Using an IBM-compatible, you could replace line 20 with:

```
20 LET R.FREQ = INP (&H3FC)
```

This would allow R.FREQ to take on a value from data fed through an external serial port (COM1). You could also output data through a serial port by using an OUT statement in line 40.

Suppose we want to use calculations which deal with values expressed in hertz. In the first example, we would only need to append three 0s to each constant. The BASIC interpreter is already set up to handle larger numbers, and no added programming steps are necessary.

Were the same task to be undertaken with an assembly language program, it probably would have been written to handle variables and constants no larger than necessary. This would allow the computer to perform the calculation as quickly as possible. Fewer resources would be used, allowing more to be allocated for other tasks.

Using larger numbers may mean the finely tuned assembly routine will have to be rewritten. Depending on the size of change and type of calculation, the rewritten code could be much larger than and different from the original.

Assembly language is very versatile because it allows the programmer to control any part of the machine that can be affected by software. The price you pay for such versatility is sometimes having to write a lot of code compared to using a high-level language.

Another popular language is C. I suppose a high-level language which can generate compact, efficient, fast-running machine code needs a small name. It is a *structured* language, which means it allows large programs to be conveniently made up of smaller programs. Unlike BASIC and some other high-level languages, C allows programs to interact freely with computer hardware.

While assembly language is tailored to meet the requirements of a computer, BASIC has features which tend to make it one of the easiest programming languages for most people to learn and use. My impression of C is that it is somewhere in between.

C and C++ are probably slightly more difficult for most people to use than BASIC. C certainly requires fewer programming steps than assembly language in many cases. Some mathematical procedures and input/output operations requiring many lines of assembly code can be accomplished in one C statement.

If you compare the machine code generated by an assembler to that generated by compiling and linking C source code, the C-generated code is often as compact and fast. Compiling and linking is a process the source code must undergo to produce an executable image—that is, machine code the computer can use.

Some versions of BASIC are interpreted. This forces the computer to repeatedly get another instruction, interpret its meaning, and act on it. This is a much slower process than having the machine-code instructions ready and properly assembled for the CPU to run through them.

The foregoing assessment is from someone who has dabbled in all three languages. One thing about teaching yourself computer programming is having an ever patient but stern instructor, the computer itself.

To compare the earlier BASIC listing to a similar listing in C, check out the following:

```
#include <iostream.h>
int main ()
{ int i_freq = 9000 ;
  int r_freq = 14000 ;
  int osc ;
   osc = r_freq - i_freq ;
   cout << osc;
   return 0;
  }
```

This listing was written with a C++ package that includes an editor, a compiler, and other tools into one integrated package. When compiled and linked on an IBM-compatible personal computer, it produces an executable file for the same system. It looks similar to the previous BASIC listing.

The first line, #include <iostream.h>, tells the compiler to add a header file to this source file before compiling. Iostream.h is needed because it allows use of the cout and cin operators described below.

The next line, int main (), defines a function. A function is a group of related instructions. The main function is where program execution begins. Functions are used as building blocks of C and C++ programs. The open brace "{" indicates the beginning of a group of program instructions.

Each group of instructions or statements ends with a closing brace "}". The semicolons ";" indicate the end of an instruction. Data items or variables such as i_freq, r_freq, and osc must be declared. This lets the compiler know what kind of data it's dealing with—in this case, an integer with a specified maximum and minimum value.

Instructions cout and cin are contained in a library of functions called iostream.h. It allows cin and cout to provide data to the program from the keyboard and from the program to the monitor screen.

C probably appears more cryptic to most people than BASIC, at least at first. All three languages are popular and useful with both personal computers and the 80C31/8051 family of microcomputers. In addition, C and BASIC can call other programs written in assembler when speed or specialized functions are needed.

So what was the transceiver software written with? It was done exclusively in assembler. The program which does this is called a *cross-assembler*. It allows you to write the software on a computing platform different from the one it is designed to run on. In this case, it was developed on an IBM-compatible PC. More on how this is done will be presented in the next chapter.

I had several reasons for choosing to write the software in assembler. The most important one was not being sure if I could accomplish some of the tasks I had in mind with another language. Part of that reason was because of my level of inexperience and part of it was because of the somewhat meager resources of the three-chip computer.

The 80C31/8051 family instruction set

An attempt at providing a reasonable tutorial on assembly language programming of the transceiver's microcontroller would require another book. Sources of such information often advertise in periodicals listed in the Addresses section at the end of the book.

Each line of assembly source code has a placeholder or field for different parts of the statement. Figure 14-1 shows a typical line of assembly code with the name for each field directly above. Some assemblers also allow a line number left of the label.

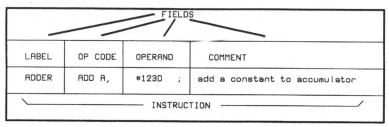

LABEL	OP CODE	OPERAND	COMMENT
ADDER	ADD A,	#123D ;	add a constant to accumulator

■ **14-1** *This is an example of adding an immediate value, a constant, to the contents of the accumulator.*

The label field lets the assembler find this line so that when it compiles other instructions which refer to ADDER, it will have the address needed by these instructions. If the other instructions, such as a jump, used a fixed address, it could become incorrect when the programmer makes changes or additions.

The source op code or operation field is a mnemonic which is usually an abbreviation for the general kind of action taken by the CPU. The exact executable machine code is determined by the assembler. In this case, adding a constant or immediate value to the accumulator will not produce the same executable binary code as adding the contents of a memory address.

Adding a constant to the accumulator requires the op code portion of a machine language instruction to be 00100100 binary. Adding the contents of a memory address requires a binary op code of 00100101 binary. With an assembler you usually don't have to worry about these details; just type ADD A,.

The operand in this case is a decimal number. Assemblers require some identification as to what number base is being used. Unlike this example, the default mode may be decimal if no identification is given. Most assemblers can accept binary, octal, hexadecimal, and decimal values.

The cross-assembler I used, as well as others, can also tell from the octothorp "#" that 123 is an immediate value. It is a constant, fixed in place at the time of assembly, and cannot be changed while the program is running.

If the characters 123D are typed alone, they refer to a RAM location. The assembler will recognize that the instruction means "add the contents of RAM location 123D to the accumulator."

Keeping track of each memory location could be difficult if you always had to use a number for identification. Another assembler trick is to use equates. An EQUATE statement can tell the assembler to equate something like the label ADDEND with RAM location 123D.

Instead of the programmer trying to remember, for example, what location he or she used for holding that variable in the addition routine, the programmer can simply type ADDEND wherever it is needed. Direct addressing with this label looks like the following:

```
ADDER ADD A, ADDEND ;
```

A semicolon ";" indicates the end of an assembly language instruction with most assemblers. Anything after this is in the comment field. If the comment requires more than one line, you can use additional semicolons.

When the operand is a constant, as in figure 14-1, the addressing mode is immediate. Modifying the example to add the contents of a RAM location illustrated how to use the direct addressing mode. Other addressing modes make use of registers located in RAM address space.

Direct register addressing, or just plain register addressing, is used in a manner similar to direct addressing. Instead of using a memory location for the operand, a register located in the RAM address space is used. The example in figure 14-1, modified to use register addressing, might look like the following:

```
ADDER ADD A, R1 ; add contents of reg. A to accumulator
```

Eight working registers, R0 through R7, are available. A register bank is made up of eight registers. The microcontroller has four banks, but only one bank can be used at a time. The selection is made by a software command.

Another addressing mode is useful when you want a variable to take on values contained in a number of different RAM locations. This is how the transceiver software accesses an array of locations which hold operating frequency data. It uses register-indirect RAM addressing.

Register-indirect addressing uses R0 or R1 as pointer registers. Their contents specify a RAM address. The address used by something like an ADD instruction can be changed by simply changing the value held in R0 or R1.

Let's use the figure 14-1 example again. We can write source code which will select values from RAM locations and add them to the accumulator. First, a value can be placed in a couple of RAM locations by the following instructions:

```
MOV 100D, #5D ;move immediate value of 5 to location 100D
MOV 101D, #7D ;move immediate value of 7 to location 101D
```

Now that RAM addresses 100D and 101D are loaded with values we want to add (for some reason) to the accumulator, we can write an instruction which will cause one of the registers to point to the first address:

```
MOV R1, #100D;move immediate value of 100 to register 1
```

If the next instruction to be executed is:

```
ADD A, @R1 ; add contents of address pointed to by R1
```

A value of 5 will be added to the accumulator because R1 pointed the ADD instruction to 100D in RAM for the operand value. Suppose you're writing a program which pages through RAM each time something like a switch closure occurs. Immediately after closure, the signal should be interrogated repeatedly to ensure that it's a valid signal. This would require using some instructions we haven't covered yet; however, assume that's already taken care of.

After going through a debouncing routine, how about having our hypothetical program add the contents from the next-higher address to the accumulator? We could use an ADD instruction to increase the value in R1 or simply increase it with an increment instruction as in:

```
INC R1 ; increase R1 by 1
```

R1 now points at a different address. When the ADD instruction is repeated, 7 is added to the accumulator. The instruction:

```
ADD A, @R1 ; add contents of address pointed to by R1
```

results in the addition of a 7 to the accumulator.

If what we are doing here seems a bit bewildering, perhaps it will help to summarize. A hypothetical program will be easier if we start with a known value in the accumulator.

```
MOV A, #0  ; move zero to the accumulator
MOV 100D, #5D ;move immediate value of 5 to location 100D
MOV 101D, #7D ;move immediate value of 7 to location 101D
MOV R1, #100D;move immediate value of 100 to register 1
ADD A, @R1 ; add contents of address pointed to by R1
; to accumulator
INC R1 ; increase R1 by 1
ADD A, @R1 ; add contents of address pointed to by R1
; to accumulator
```

The result is a value of 12 in the accumulator.

The instruction set of the 80C31 and its relatives allows most instructions to work from several different physical or logical address spaces. This is an improvement over some microcontrollers which have more exceptions or special cases.

Hopefully, this brief taste of assembly language programming of the 80C31 will give you some insight into how you could tailor the transceiver's computer program to your own liking.

It takes a lot of steps to accomplish some tasks in assembly language, but it is in many ways less mysterious than high-level languages. With a thoroughly documented instruction set and assembler, you don't have to guess what the computer is doing.

Although high-level languages may be more convenient for some tasks, the details of how the computer resources are being used may be hidden. The performance of time-critical tasks such as the synthesizer fine-tune loop may be more difficult to predict.

Input/output and other special functions

As we saw in earlier chapters, the 80C31 has 32 input/output pins. These are configured as four 8-bit parallel ports named P0, P1, P2, and P3. Each port can input or output information under the command of a number of different software instructions. The port pins can be used one at a time to handle single-bit I/O in addition to conventional bytewide operations.

The parallel I/O ports have addresses in a range occupied by the special functions register. The list of special functions registers includes those related to the timer/counter logic, serial communications, interrupts power management, and other tasks.

The instruction to output a value on port 1 could be as simple as:

```
MOV P1, #15D; move 15 to port 1
```

The binary value would be 00001111. The rightmost 1 appears as a high at pin 1. The leftmost 0 appears as a low at pin 8.

For the previous assembler instruction to work, it must know the address of P1. This may be accomplished with an *equate directive*. Using P1 as an example, its address is 90H (hexadecimal) or 144 as a decimal value.

Other special functions registers are perhaps a bit more complicated to use because each bit in the register has a distinct function. Consider again the capabilities of timer/counter logic in the 80C31 with figure 10-8.

Timer/counter mode control register TMOD affects how the timer/counter logic is configured. The control bits are arranged in the manner depicted in figure 14-2. Control bits for timer 1 behave the same as timer 0 bits.

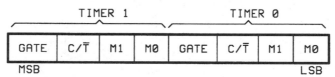

■ **14-2** *The timer/counter mode control register, TMOD, is one of about 20 (depending on the microcontroller version) special functions registers.*

This description of timer 0 also applies to timer 1. The timer-0 gate bit can be used to turn the pulse train on and off if the timer-0 run control bit in another register called TMOD is high and the interrupt-0 pin is low.

If GATE is high, the interrupt-0 pin can control the pulse train. If GATE is low, timer-0 run control in TMOD can control the pulse train. A bit called C/not-T controls whether CPU oscillator pulses or external pulses are counted.

M1 and M0 control the operating mode of the counters. When they are both zero, timer/counter 0 is operating as a 13-bit timer/counter, as shown in figure 10-8. If M0 is changed to 1, 16 bits can be counted before rollover.

The timer/counter registers can use three other modes, but I won't go into that here because they are not used in the transceiver. The main point is that by studying information from tutorials, programming text, and manufacturers, you can customize the original transceiver software or make your own application.

An overview of 80C31 instructions

Most instructions use a variety of addressing modes. As we observed earlier, the MOV instruction has 15 variations when addressing modes combine with various sources and destinations for the data being moved. Rather than go over each addressing mode or variation of each instruction, let's examine the general case of each instruction.

We have already seen a couple of the arithmetic operations. These were the instructions ADD and INC. Another addition instruction is ADDC, meaning "add with carry."

ADDC is one of several instructions which make use of a bit called the *carry flag*. The carry flag is one of several bits in a special register known as the *program status word* (PSW), which is used by various instructions to make decisions.

One of the conditions which sets the carry flag is an overflow from the addition of two unsigned integer values resulting in a sum greater than 255. ADDC simultaneously adds the byte variable, the carry flag, and the accumulator contents. The result is left in the accumulator.

SUBB is the mnemonic for "subtract with borrow." It is subtraction's version of ADDC. SUBB sets the carry (same as borrow) flag if a borrow is needed for the most significant bit. If a borrow is not needed, it clears the carry bit.

DEC is the opposite of INC. Neither DEC nor INC affects the carry flag. Along with affecting the accumulator, they can also be used with register addressing, direct addressing, and register-indirect addressing.

INC DPTR is the mnemonic for "increment the data pointer." It is useful when the program needs to get data from a table consisting of a series of entries in ROM or EPROM.

The 16-bit-wide data pointer serves as a base register in indirect jumps, table look-up instructions, and external data transfers. Upper and lower 8-bit halves of the data pointer may be manipulated separately or together.

MUL AB multiplies the unsigned 8-bit contents of the accumulator and register B. The low-order byte of the 16-bit product is left in the accumulator. The high-order byte is left in register B. Another flag in the PSW register, called the *overflow flag*, is set if the product is larger than 255.

DIV AB divides the unsigned 8-bit contents of the accumulator by the unsigned 8-bit contents of register B. The accumulator receives the quotient. The remainder is left in register B. If division by 0 is attempted, the accumulator contents will be undefined and the overflow flag will be set.

DA A is an instruction which is used after the ADD or ADDC instruction is performed on a packed BCD number in the accumulator. The packed BCD format contains two BCD numbers in one byte. DA A rearranges the resulting sum to obtain a correct value.

Another group of frequently used instructions is *logical operations*. ANL is the logical AND operation for byte variables. The operation byte and operand byte will be compared bit for bit. When both bits of the same place value are 1, the same bit in the operation (destination) byte will be a 1. If they are not, the resulting bit is 0.

```
MOV A, #11110000B;
ANL A, #10101010B;
```

leaves 10100000-binary in the accumulator.

ORL performs a bit-by-bit logical OR operation between the destination byte and source byte, storing the result in the destination byte. If either bit of the same place value is 1, a 1 is left in the same position of the destination byte.

```
MOV A, #11000000B;
ORL A, #10101010B;
```

leaves 11101010-binary in the accumulator.

XRL is the exclusive-OR mnemonic. XRL performs a bit-by-bit logical exclusive-OR operation between source and destination bytes. If one bit of the same place value in either byte is 1, a 1 is left in the same position of the destination byte. If both bits are 1, the destination bit is 0. If both bits are 0, the destination bit is also 0.

CLR A resets all accumulator bits to 0. No flags are affected. CPL A reverses, or complements, all accumulator bits, and no flags are affected.

RR A means all accumulator bits are rotated 1 bit to the right. The least significant bit is rotated into the most-significant-bit position. No flags are affected. See the following example:

```
MOV A, #00000010B; put 2 in accumulator
RR A; this instruction will change accumulator to 1
RR A; this instruction will change accumulator to 128D
RR A; this instruction will change accumulator to 64D
```

RL A rotates the accumulator bits one position to the left. The most significant bit is rotated into the least-significant-bit position. No flags are affected. Here is another example:

```
MOV A, #01000000B; put 64D in accumulator
RL A; this instruction will change accumulator to 128D
RL A; this instruction will change accumulator to 1
RL A; this instruction will change accumulator to 2
```

RRC A rotates the accumulator and carry flag 1 bit to the right. The least significant bit moves its value into the carry flag. The original value of the carry flag moves into the most-significant-bit position.

RLC A rotates the accumulator and carry flag 1 bit to the left. The most significant bit moves its value into the carry flag. The original value of the carry flag moves into the least-significant-bit position.

SWAP A swaps nibbles within the accumulator. The 4 most significant bits exchange places with the 4 least significant bits. If the accumulator contains 10101111B, the execution of SWAP A leaves 11111010B in the accumulator. This is the equivalent of a 4-bit rotation.

Data transfer instructions comprise a group of frequently used instructions. The MOV mnemonic is used for most instructions in this category. It is very versatile and allows the program to move almost any kind of byte data to almost any bytewide destination.

In addition to the previous examples, you will see plenty of instructions in the main program listing (see appendix A) which use MOV.

MOVC A, @A+PC BASE_REG is used to load the accumulator with a code byte or constant from program memory. The address of the byte fetched is the sum of the original unsigned 8-bit accumulator contents and the contents of a 16-bit base register. The base register, designated here by BASE_REG, can be the data pointer or the program counter.

To load the data pointer with a value so that it can point to a location in program memory, use the instruction MOV DPTR, CONSTANT, where constant is a 16-bit immediate value which has been previously given a numerical value by an equate directive. An example with an numerical operand might look like MOV DPTR, #1234H;.

MOVX instructions all deal with external RAM. They work much like the other MOV instructions. No external RAM is used in the transceiver's computer.

PUSH and POP are instructions which input and output bytewide data to a special area of RAM called the *stack*. The stack is especially useful for temporarily storing variables when the normal progress of the program is interrupted.

PUSH VARIABLE, assuming VARIABLE has been given a value, does two things. It increments the stack pointer, which is a register for keeping track of where stack operations will occur. It also stores the contents of VARIABLE in the new RAM address.

Pushing a register or the accumulator onto the stack is often done to save it before some interrupt routine takes place which could change it and wreck whatever the program was doing before the interrupt occurred.

Several PUSH instructions can be executed to save any necessary data. After the routine triggered by an interrupt is completed, the stack pointer contains a value which indicates the address-holding value of the last value saved.

POP can be used to read each value back into its corresponding register or variable. When the value is read from the stack, the stack pointer is decremented. This process reads the several values saved by PUSH, starting with the last one first.

XCH A, exchanges the accumulator with a byte variable. Accumulator contents can be exchanged with the contents of a RAM address, a register, or an address pointed to by a register. This corresponds to direct, register, and register indirect addressing.

XCHD A, @Ri, where "i" can be 0, 1, 2, or 3, exchanges the least significant nibble of the accumulator with that of the RAM location indirectly addressed by the specified register. The most significant nibble of each register is not affected.

An unusual and effective capability of the 80C31/8051 family is its ability to manipulate single-bit variables without resorting to bytewide operations to mask other bits. Such single-bit instructions are classified as *boolean variable manipulations*.

Instructions which rely on the carry flag include SETB C for "set the carry flag." CLR C means "clear the carry flag." CPL C means "complement the carry flag."

The carry flag can be logically ANDed with BIT by the instruction ANL C, BIT. A section of RAM is bit addressable. Each individual bit can be specified by its own individual address. In this case, an equate directive could be used to assign one of those addresses to the variable, BIT.

ORL C, BIT works in the same manner as the previous bytewide example, except that the bits are ORed together. MOV C instructions are similar to the earlier bytewide MOV instructions, except a bit address is used instead of a byte address.

All of these boolean manipulations can also be written with a bit variable in the operand instead of the carry flag. Assuming the bit address of port 1.1 equates to P1.1, a single port pin could be toggled on and off by:

```
SETB P1.1 ; pin 2 (second pin of port 1) is high
CLR P1.1 ; pin 2 (second pin of port 1) is low
```

Not only must data be moved and manipulated, the computer needs instructions for changing the course of action as conditions warrant. A variety of branching and control instructions is available in the 80C31/8051 instruction set.

ACALL is an *absolute subroutine call*. It unconditionally calls a subroutine located at the indicated address. It increments the program counter twice to obtain the address of the following instruction. The 16-bit address value is pushed onto the stack, incrementing the stack pointer twice.

ACALL is limited to calling subroutines no more than 2 KB away from the first byte of the instruction which follows. LCALL is similar, but it can call subroutines located anywhere in the 64-KB program.

RET is the mnemonic for "return from subroutine." It pops the high-order byte and then the low-order byte from the stack. This action decrements the stack pointer by 2. Program execution continues at the resulting address, usually the instruction immediately following an ACALL or LCALL.

RETI is used when program execution has returned from an interrupt. RETI pops the high- and low-order bytes of the program counter from the stack. The stack pointer is decremented by 2 and program execution continues at the resulting address, usually the instruction immediately following an ACALL or LCALL.

SJMP, AJMP, and LJMP are jump instructions where the program execution branches unconditionally to an address determined at assembly time. SJMP has a range of 128 bytes, AJMP can span 2 KB, and LJMP allows branching anywhere in the 64-KB address space.

JMP @A+DPTR means "jump indirect." It adds the data pointer contents to the unsigned contents of the accumulator and loads the resulting sum into the program counter. It allows the program to jump to an address which is not fixed at assembly time. The jump destination can be changed while the program runs.

Other branching instructions are not unconditional. They test for some value or condition and make the branching action dependent upon the test results. Most conditional branch instructions have a range limited to 128 bytes.

JZ means "jump if the accumulator is zero." JNZ means "jump if the accumulator is not zero." JC means "jump if the carry flag is set." JNC means "jump if the carry flag is not set."

Bit values can be used to make conditional branching decisions. JB means "jump if the direct bit is set." JNB means "jump if the direct bit is not set."

CJNE stands for "compare and jump if not equal." It compares the magnitude of two operands and branches if their magnitudes are not equal. A hypothetical instruction might compare the contents of R7 with an immediate value of 123D and branch if they are not equal to an address in program memory labeled SKIP.

```
CJNE R7, #123D, SKIP
```

CJNE allows four addressing-mode combinations with the first two operands. The accumulator may be compared with any directly addressed byte or immediate data. Any indirect RAM location or working register may be compared with an immediate constant.

DJNZ stands for "decrement and jump if not zero." A 1-byte address is specified in the op code and its contents are decremented by 1. A DJNZ instruction which decrements register 1, then asks if it's zero, to decide whether or not to jump to a location labeled TH_PLACE, would look like:

```
DJNZ R1, TH_PLACE
```

NOP is an easy instruction to describe. It does nothing but use up a CPU cycle. Only the program counter is affected and advances one count. It is sometimes intentionally used to waste time to allow some external event to catch up to the program's execution.

Where to go from here

The preceding abbreviated view of the 80C31/8051 instruction set does not have the degree of detail necessary to write extensive revisions of the transceiver's source code. Inexperienced experimenters may find the information in this book sufficient for simple modifications.

To acquire a better understanding of the instruction set and microcontroller hardware, additional sources should be consulted. Intel's *Embedded Applications* and *Embedded Microcontrollers* (see Bibliography) or similar texts should be included in your arsenal if you plan to mount an assault on problems encountered in a particularly ambitious project.

Many hardware details have not been covered here. Some nuances of the microcontroller hardware can be deduced by the more astute software experts who can reverse-engineer portions of the program. It is a lot easier to look up information in readily available texts and is certainly advisable.

If you have little or no computer experience but a strong desire to experiment with the firmware/software in this project, don't be discouraged. You will need to learn some programming basics and have access to an IBM-compatible personal computer and some programs used as tools for the task. Part of the next chapter will discuss some of these tools.

Developing
the firmware

15

WHEN THE LONG LIST OF MACHINE INSTRUCTIONS IS finally stored in stable memory such as EPROM, I suppose it's correct to call it *firmware*. Supposedly, it is embedded in the transceiver's computer for the life of the radio, never to be changed again.

The foregoing assumption may, in some cases, be correct. We know from experience that many people like to customize their projects to accommodate their personal views of how various features may be included or modified in a given project.

Experimenting with the firmware/software will require the builder to go through some of the same steps that were used to develop the program in the first place. You have the advantage of not starting out from scratch.

With a blank memory space, I couldn't even input or output data to see what is happening. A fair amount of code had to be written before I could even tell if anything was working.

On the other hand, the existing program has complexities which must be understood to make extensive modifications. It's the old chain-reaction effect, where you change one thing and that changes something else, and so on.

Unfortunately, the effect of some of those reactions is simply to crash the program, making it difficult to debug the problem. Sometimes, some rather sneaky tests or traps must be laid to catch the condition which causes the malfunction.

Regardless of how simple or how complex the software, we must decide what it is supposed to do before we write it. This may sound a bit simpleminded, but the requirements are not always as clearly defined as we might first imagine.

Some uncertainty existed when the system's input and output data depended on radio hardware with characteristics that were not completely known until the entire system was up and running.

An overview of software tasks

All of the comments in this section refer to various portions of the Main Program Listing (see appendix A). You can refer to the listing by thumbing back and forth as various parts are discussed. An even better solution might be to have a copy displayed on your computer monitor.

The first part of the assembly source code replaces a long list of numbers with names. Equate directives are used to give names to many constants, variables, and RAM addresses, which become part of the machine language instructions in the firmware.

As an example, the first few lines look like this:

```
;MCS-51 INTERNAL REGISTERS
;
B:      EQU     ØFØH        ;B REGISTER
ACC:    EQU     ØEØH        ;ACCUMULATOR
PSW:    EQU     ØDØH        ;PROGRAM STATUS WORD
IPC:    EQU     ØB8H        ;INTERRUPT PRIORITY
P3:     EQU     ØBØH        ;PORT 3
```

Some assemblers or cross-assemblers for the 80C31 may not need to have an equate table such as this. The cross-assembler I used works with many different types and brands of microcontrollers, so these assignments are necessary.

Because a number of instruction tables are available for different processors, another directive,

```
CPU "8Ø51.TBL"
```

chooses the 8051 instruction set. This happens on the 75th page line (this is not the 75th instruction that occurs in the program), just a couple of lines before the first op code.

The next directive, HEX "ON", turns on the hexadecimal output file. Data from the hexadecimal file is turned into machine language by other hardware and software involved in transferring it from the personal computer to memory in the 80C31 three-chip computer.

Line 77 has a directive which allows the user to specify the value of the program counter during assembly. The first address in the program counter is 00.

224

Let the instructions begin

AJMP 100H at line 78 is the first line containing a real machine instruction. The reason for jumping from 00 to 100H is that a couple of interrupt routines have their starting addresses within this area.

When an external interrupt occurs, an instruction built into the 80C31 hardware sends program execution to a predetermined address. In the case of interrupt 0, the starting address is 03H. Interrupt 1 uses 13H.

If the routine triggered by an interrupt is small enough, it can be written to use the address space below 100H. If it will not fit, program execution can be vectored off to some other address by a jump instruction as in

```
LJMP COUNT
```

in line 80. COUNT is a routine, which we will examine later, that tallies the number of pulses arriving at the interrupt-0 pin over a period of time.

What this means is that the meat of the program, where some useful processing is performed, really starts at 0100H. The following is a summary of the major tasks performed by the program's main loop:

> At 100H, start the display initialization and setup.
>
> Load reference-counter shift register in PLL chip.
>
> Set up timers and interrupts.
>
> Define a bunch of variables with equate directives.
>
> Start the main polling loop.
>
> Check for power failure; implement power-down routine?
>
> Test frequency memory latch; select band-switched filter.
>
> Select DAC and PLL inputs depending on RX or TX mode.
>
> Display opto-counter/DAC value if in diagnostic mode.
>
> Manipulate frequency memories.
>
> Display letter name of frequency memories.
>
> Test for split-frequency modes; set tuning capability as needed.
>
> Calculate intended output frequency of synthesizer mixer.
>
> If diagnostic mode, display intended output frequency of synthesizer mixer.

Display emission mode.

Extract PLL divisor.

Calculate actual output frequency of synthesizer mixer.

Display split-frequency selection.

If split-frequency mode, manipulate DAC input value.

Compare actual to intended mixer output frequency and calculate DAC value.

Load A and N counter of PLL.

Debounce and read push buttons.

Select tuning rate of rotary shaft encoder knob.

Read opto-coupler values of rotary shaft encoder.

Select and store emission modes.

These are most of the major functions handled by the main polling loop. They are visited over and over again in an endless loop. Some variables are polled, or interrogated, each time through. Others are updated when specific events occur.

Subroutines

Outside the main loop exist a number of subroutines ready to be called only when they are needed. Some of these subroutines—or *procedures*, to use a more modern and less descriptive term—are called several times through each loop. Others are called only occasionally.

Copies of subroutines called in every cycle of the main loop could be included in the main loop. It would allow the program to run slightly faster; however, the source code would probably be even harder to follow than it already is.

The following is a summary of most of the tasks performed by the subroutines outside the main loop:

Binary to binary-coded decimal conversion.

BCD to ASCII conversion.

Position display characters.

Read and write to display.

Measure synthesizer-mixer output pulses.

40-bit divide subroutine.

40-bit multiply subroutine.

32-bit add subroutine.

32-bit subtract subroutine.

Reload DAC and PLL if transmitting in split-frequency mode.

With lists like these, describing every program function in detail in this chapter is not practical. We can examine some of the more interesting and important examples.

Getting started

Various parts of the program were developed as various hardware was brought online. The first step was to try and get the three-chip computer up and running. The only input/output devices at this point were some push-button switches and an oscilloscope.

With nothing in program memory, the system does absolutely nothing. It's not like a personal computer that can at least accept operating-system software because the basic input/output system (BIOS) is in place.

After confirming that a small section of code could actually read and toggle port pins, communicating with the display module was the next challenge. Another struggle ensued before I had code that would initialize the display and communicate correctly over a 4-bit bus.

Having a functioning display was a great boon to developing the rest of the firmware. Before I could capitalize on the display capabilities, conversion routines had to be written which would allow displaying decimal numbers.

Being no software expert, I decided to design the main program to crunch binary numbers. The end result of calculations to be displayed is converted to BCD then ASCII. Input data from the push-button switches and rotary shaft encoder is already in binary format and needs no conversion.

The rest of the system development proceeded with a bit of hardware added here, some more firmware there, etc. The first radio circuit to be interfaced with the three-chip computer was the frequency synthesizer. The remainder of the radio hardware was then completed.

Highlights of low-level code

I suppose the best place to start describing selected parts of the Main Program Listing (see appendix A) is where it says "START" at the 85th line. P3.4 and P3.6 are cleared. Port pins P3.4 through

P3.7 are manipulated throughout the main loop to enable or disable hardware wherever necessary.

After the necessary delays are implemented by looping, some commands are used to set the display in the proper mode. With the display initialized, the PLL reference counter value is sent. This value (600 decimal) is constant and needs to be sent only once when the transceiver powers up.

The timer and external interrupt modes are set by simply putting the correct bit or byte values in the appropriate registers. Intel and other reference sources contain detailed information about how to manipulate these registers.

Starting at the main loop on line 339 is an interesting bit of code. If supply power fails, P3.7 is low. A low at P3.7 causes the routine to write lows to ports 1, 2, and 3. A value of 2 written to the power control register, PCON, stops operation and puts the microcontroller into power-down mode.

Another major section of code, starting around line 349, tests the value in the active frequency memory location. A 4-bit value is selected for writing to the 4094 shift register. This value ultimately is decoded to select one of the low-pass bandswitched filters connected to the antenna.

If the operating frequency is above the top edge of a particular amateur band, the next-higher filter is selected. For example if you are tuned to 3999 kHz, the 75/80-m filter is selected. When the transceiver is tuned out of the band to 4001 kHz, the 40-m filter will be selected.

Another test is performed which checks to see if the operating frequency is below the bottom band edge, along with checking for transmit mode. If the operating frequency is not within an amateur band and transmit mode is engaged, no filters are selected.

With no filters in line, the rig can't get a signal to the antenna. It's a brute-force method of guarding against transmitting outside amateur bands. You can change the band-edge limits if you find it necessary. Convert the new limit and then unpack it byte by byte, replacing the old values.

Generating synthesizer data

Starting around line 559 is a comment that states, "D TO A AND PLL MEMORY SELECTION." The DAC and PLL receive their digital inputs from the contents of different registers, depending on whether the frequency in use is tunable or fixed.

In order to use every last byte of RAM, FUNC has two variables packed into it. The split-frequency mode is designated by the low-order nibble and the tuning rate is packed into the high-order nibble to be used in a different routine.

Selection and writing of the PLL and DAC values is one of the more convoluted parts of the source code. If you are trying to decipher this or other parts of the code and wondering what in the world is going on here, don't feel alone. I sometimes do the same thing, and I wrote the stuff!

One useful source of information is the equate statements. Many have a brief description of the variable involved. Most of the labels, especially jump destinations, are shortened forms of a word or phrase to describe why they are there. RXMD is, for example, associated with receive (RX) mode.

An essential value for the synthesizer is the number used for dividing the VCO frequency in order to match the reference frequency. A routine starting around line 750 extracts this divisor from the value of the operating frequency.

Finding the divisor's value is essentially the act of ignoring the lower 13 bits of the frequency memory because this lower-order part of the frequency will be accommodated by the fine-tune loop of the synthesizer.

The fine-tune loop and associated frequencies

This is one of those places where having a binary value for the top frequency of the reference oscillator is an advantage. A multiple-byte division routine would be larger and have a much longer execution time.

In chapter 10, we saw the formula for calculating the mixer output frequency. At line 800, RAM locations are loaded to begin the calculations. The variables named CTRLO and CTRHI contain the high and low bytes of the total number of timer pulses since the last reset.

An unnamed RAM location, 2BH, contains the number of mixer output pulses since the last reset. CTRLO, CTRHI, and location 2BH all receive data in a subroutine called COUNT. COUNT is called each time a hardware interrupt is triggered by pin 2 on port

3 going from high to low. COUNT effectively measures the length of a pulse in units of timer cycles.

The number displayed as the operating frequency uses the following constants and variables:

MX_{out} = synthesizer mixer output frequency in Hz.

CPU_{osc} = 9,831,396 = CPU clock oscillator frequency in Hz.

IF = first intermediate frequency in Hz. For SSB modes, it represents the value necessary to make the display indicate where the missing carrier would be. For CW, the display should indicate the center of an emitted signal.

PLL_d = phase-locked loop divisor.

N_r = 600 = factor by which voltage-controlled reference oscillator is divided to equal reference frequency.

OP_{freq} = operating frequency in hertz.

The terms relate to each other in the following equation:

$$OP_{freq} = PLL_d \left(\frac{CPU_{osc} - MX_{out}}{N_r} \right) - IF \qquad (15\text{-}1)$$

We could run through an example using actual numbers. Let's say MX_{out} = 1600 Hz, PLL_d = 5251, and IF = 72,000,000 Hz. The operating frequency is 14,027,098 Hz. Ten hertz is the smallest resolution, so this would be an example in which the system tried to hit 1,427,100 and missed by 2 Hz.

It is not a system in which the operating frequency is precisely what the display indicates. The amount of error is more than small enough to cause no problems with SSB and CW modes. Stability and accuracy will depend, of course, on the CPU clock oscillator, which acts as the master frequency-determining element of the whole system.

Another interesting portion of code essential to the synthesizer starts at line 914. This is where the intended and actual frequency of the fine-tune loop's mixer output are compared to arrive at a value for the DAC. This is also probably one of the better places for experimentation because it's important to how the fine-tune loop operates and there is probably room for improvement. The first few lines compare the mixer output frequency to a variable named CALM.

CALM is equal to the least significant byte of the last valid mixer output frequency. If CALM and the least significant byte of the present frequency are equal, no action is taken because execution jumps to NOOMO, skipping the instructions that generate DAC values.

If CALM is different from the present least byte, the routine will test to see if the mixer output frequency is higher or lower and decides whether to increase or decrease the DAC value. The DAC value is proportional to the frequency difference. Rotate instructions are used to halve the DAC input value. Subtraction or other manipulations could be tried here.

An ideal situation would be one in which the DAC input value would exactly cancel the frequency error on the first try. The fine-tune loop could then settle on frequency consistently in the least possible time. For this to happen, the crystal-controlled VXO would need to have a linear voltage-to-frequency relationship.

The simplest relationship might be one in which each bit at the DAC input translated to a 1-Hz change at the oscillator. Presently, the voltage-to-frequency transfer function of the crystal-controlled VCO is nonlinear and must be compensated by the above-mentioned divide-by-two method.

Loading the auxiliary and main (A and N) counters of the PLL starts at the 988th line. This is one of the few places where external interrupt 1 is not allowed to occur because the serial data stream can be corrupted. After external interrupt 1 is temporarily disabled, values from PLMH (PLL memory high) and PLML (PLL memory low) are shifted into the PLL chip.

Selection of the A and N counters is accomplished by the 18th bit. If P3.0 is high, making it a 1, the R (reference) counter will be selected instead. Because the PLL shift register and the 4094 interface shift register are chained together, they must be treated as one long 26-bit register. As evidenced by P3.6, the DAC can be loaded separately.

Reading input hardware and configuring output data

Continuing through the program, we find the push-button reading routine at line 1179. Push-button memory (PBM) stores the value of each button-bit only after at least one of the buttons has been pressed for a length of time determined by constants in the routine labeled TIMEO.

Some parts of the program use PBM for making decisions. Other routines, such as TUNING RATE SELECTION at line 1211, use TMP_0 to determine the momentary status of push buttons on the

P1 bus. A 2-bit value packed into FUNC determines the tuning rate. If you want to add more tuning rates, the instruction at line 1229 stating ANL A, #00110000B ; LIMIT UPPER NIBBLE TO MODULO 4 could be changed to ANL A, #01110000B ; LIMIT UPPER NIBBLE TO MODULO 8.

New selections to the code following this instruction could subsequently be added. Values in TMP_1, TMP_2, and TMP_3 govern how much the rotary shaft encoder will increment or decrement its variables.

Reading the rotary shaft encoder's optically interrupted phototransistors should be interesting because we've already discussed the flowchart in figure 13-4 describing how it works. A 3-byte quantity corresponding to the operating frequency is held in R5, R6, and OPTHI. It is later loaded into a location pointed to by FMML (frequency memory location) when the main loop is completed.

Another section of code where TMP_0 communicates the pushbutton status is the SAVE MODE routine at line 1311. MODE refers to emission mode, not to be confused with split-frequency modes discussed elsewhere. ASCII codes are sent to display the proper characters and the mode bits are stored in the upper nibble of SHRG (shift register).

At the end of SAVE MODE is the instruction LJMP POLL. This is the end of the main loop. Everything after this point is called as a subroutine. MAKE BCD at line 1389 converts a 3-byte binary value contained in BIN, BIN2, and BIN3 to an 8-digit packed BCD format contained in BCD, BCD2, BCD3, and BCD4.

Packed BCD is a step in the right direction but must be converted to ASCII codes that the display module can understand. DISPLAY DECIMAL NUMERALS at line 1429 performs this conversion. To do so, it calls another subroutine at SETAD, which sets the starting position for the characters.

Each time another character is needed, WRITE is called and the module automatically moves the next character's position to the left. Both SETAD and WRITE call READ to check the module's busy flag before proceeding.

Interrupt and math subroutines

COUNT at line 1522 is the destination of an LJMP instruction vectored to by external interrupt 0. It has its own counting routine, which allows the hardware timer/counter to count as many timer pulses as possible before coming too close to rollover. Each counting period is tallied by RAM location 2AH.

Notice that the program status word, accumulator, register B, and data pointer are pushed onto the stack. This interrupt could occur in the middle of almost anything and will disrupt the program's execution if vital information is not restored. The above-mentioned registers are pushed off the stack in reverse order before returning to an interrupted procedure.

Next comes divide, multiply, add, and subtract integer routines adapted from examples in Intel's *Embedded Applications* (1990, see Bibliography). The 40-bit multiply and divide routines may at first seem larger than necessary for the numbers used here. Large numbers are encountered because some terms must first be multiplied before division can take place to eliminate losing bits as remainders.

The last section of code is active when using split-frequency modes. Triggered by external interrupt 1, it reloads the PLL and DAC with new values when a frequency change is needed. Other related parts of the program have already saved the DAC input value so that when you hit the key, the program stops what it's doing—that is, unless serial communication is taking place.

External interrupt 1 is disabled during serial communication periods because the shift registers end up with corrupted data if the serial data stream is interrupted partway through to start a new one. Fortunately, these time periods are short enough to allow a fast response even suitable for full break-in CW operation.

More program development tools

For writing the source code, the cross-assembler was obviously very useful. It requires an ASCII editor. Working in a DOS environment, I found a terminate and stay resident (TSR) program, called Sidekick from Borland, useful.

The cross-assembler I used is made by Universal Cross Assemblers. To get the cross-assembler's output file from the personal computer's disk drive to the 80C31 program memory requires additional hardware and software.

An EPROM programmer with matching software can take the assembler's output file and program machine-language instructions into an EPROM. The EPROM programmer I used for this project is from Needham's Electronics.

With this minimum amount of equipment and an EPROM eraser, you can modify and experiment with the program. However, I did not develop the program in this manner. With all of the lessons learned by trial and error, I would probably still be trying to get the display to work.

Instead, I used another gadget which impersonates an EPROM, called an EPROM emulator. The emulator I used is from Parallax Inc. Its connector plugs into the EPROM socket, temporarily replacing the real EPROM. A hardware and software interface converts and downloads the cross-assembler's output file to static RAM.

The RAM stores and accesses instructions just as the EPROM would. To try a program change, all you have to do is download the new file. It is not necessary to pull the EPROM, erase it, reprogram-program it, and plug it back in. Figure 15-1 represents how some systems may look.

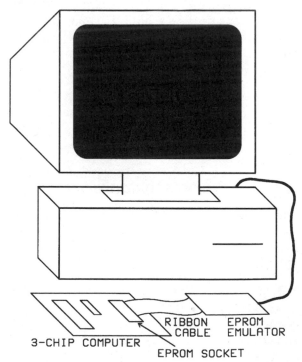

■ **15-1** *A setup for program development might look something like this. The three-chip computer in a real situation will probably be connected to the display module, synthesizer, and other circuits.*

Some EPROM emulators attach to the desktop's parallel port; others use a serial port or even an expansion slot. Some use a short ribbon cable to connect to the EPROM socket, while other small units have a rigidly mounted plug, allowing you to plug the whole emulator into the socket.

When everything is set up correctly, you can make a change in source code, assemble it, create an output file, and download it to the 80C31 system in seconds. This is much more convenient than trying the same change with EPROMS, which would take minutes.

You may be able to automate the development process by using a small batch file with an IBM-compatible similar to this example, named AS.BAT:

```
ECHO OFF
C16 NOTES -H OUT
TROM OUT
```

The ECHO OFF command is not really necessary if you don't mind the small amount of screen space taken up by the batch-file commands. The second line uses the C16 command to tell the cross-assembler to assemble instructions in a text file named NOTES and produce a hexadecimal output file named OUT. TROM is a command which tells the EPROM emulator interface to send information in the OUT file to the EPROM emulator.

If a connection is made between the EPROM emulator's auto-reset output and the 80C31 reset pin, you need only type AS and hit the enter key to try a change in source code. These commands are specific to the particular tools which I used for this project and do not necessarily apply to other software and equipment.

Using an EPROM emulator is certainly not the only method of developing software for the 80C31. Hardware is also available which emulates the microcontroller itself. Most in-circuit emulators have many features designed to facilitate software development.

By monitoring and communicating with the target system, an in-circuit microcontroller emulator can show details of the program's operation not directly observable with an EPROM emulator. Debugging some programs (finding mistakes or flaws) is, in some cases, easier if you can see how the program is proceeding step by step with a trace operation.

Setting conditional or unconditional breakpoints to stop normal-speed execution at a selected point can also be useful. Single-step operation, which allows you to manually trigger the operation of a single instruction, might help zero in on a program bug.

Other nice features are also available. The only problem is that many in-circuit emulators will cost as much or more than the transceiver. That may be a little hard to justify unless you use it in your work or plan to build a lot of microcontroller-based projects.

Troubleshooting bugs

So much for the bad news. The good news is that a number of tactics can be employed to help debug code with only an EPROM emulator. It certainly helps to have a versatile display like the module used in this project.

Other peripheral devices, such as the 4094 shift register or the DAC, have outputs which can be measured with a meter or other instruments to see what's happening. In a few instances, it was not possible to evaluate the effect of a software change until the whole transceiver was up and running; however, this is the exception rather than the rule.

Relatively inexpensive microcontroller simulation software that runs entirely on a personal computer is another way to gain familiarity with the 80C31 and many other microcontrollers. Although a pure simulator doesn't interface with any hardware, it could be a valuable way to gain experience with programming the microcontroller and test new ideas.

A simple but helpful tactic to keep in mind is to make any changes in the program as small as possible and evaluate each change as it is implemented. Changing a lot of code at once is an easy way to get into trouble.

Of course, it's not always possible to make small changes. If you need to add another multiple-precision math routine, a certain number of instructions are required before it will work at all.

If it is a routine that does not need extensive interaction with the main program, you might be able to assemble it as a small, self-contained program, complete with trial input values and see if it runs correctly by itself before stuffing it into the main program.

When things weren't working out well on a section of code, careful observation of the three-chip computer and trying small changes often provided the answer. At other times, it became obvious I was trying to do something the wrong way and had to backtrack to come up with a different idea.

A variety of techniques are useful for finding bugs. You can temporarily include diagnostic sections of code which test values in suspected registers or variables and displays them. In addition to ending up with the wrong value somewhere along the line, some bugs cause program flow to take a wrong turn, executing instructions you didn't expect.

Temporarily eliminating or changing a suspected conditional branch instruction may help locate a problem, assuming you can observe the effects. If the bug causes a crash, or the values change too rapidly to be observed on the display, or a number of other conditions occur, you may not be able to glean much information from the behavior of the system. In such instances, you may find it helpful to temporarily freeze execution at selected points.

Freezing program execution can, in some cases, be as simple as the following instruction:

```
STOP: SJMP STOP
```

An endless loop ensues, which allows everything else except interrupt-generated routines to hold a steady state. If you also want to stop any routines triggered by interrupts, you could include something like:

```
MOV IEC, #0 ; turn off all interrupts
```

before the STOP label—that is, assuming an equate directive has already assigned 0A8H, the interrupt enable register, to IEC.

Another debugging possibility is to implement single-step operation by using external interrupt 1. In general, you could also use external interrupt 0. With this project, external interrupt 0 is already in use as the frequency-determining input of the fine-tune. External interrupt 1 is also in use as a transmit/receive monitor. It can be disconnected with less disruption to the transceiver's operation than interrupt 0.

To use P3.3 as a single-step command input, TCON, the timer/counter control register must have the four's-place (third most significant) bit cleared. This causes interrupt 1 to be triggered by a low logic state instead of by a high-to-low transition.

To change the bit in TCON, the whole register can be addressed directly at 088H, or the bit itself can be addressed at 08AH.

The following routine should be placed in program memory at 013H:

```
JNB P3.3, $;wait here until external interrupt 0 goes ; high
JB P3.3, $;now, wait here until it goes low
RETI ; go back and execute one instruction
```

This would replace the current instruction LJMP CHFAS. The "$" character is interpreted by the assembler as the current value of the program counter.

In operation, most of the time will be spent looping in the second instruction. When P3.3 goes low, RETI returns operation to the next instruction of the main program. Only a single instruction gets to execute. As soon as it's done, interrupt 1 vectors operation to the first instruction in our single-step routine.

To make this scheme work in a reliable manner, you will probably need to drive P3.3 with a one-shot flip-flop or some device that can generate a clean, debounced pulse. Otherwise, the press of a switch may cause several instructions to execute.

238

Transceiver design choices

16

SHIELDED CIRCUITS ARE USED IN RADIO APPLICATIONS for a variety of reasons. Even the least expensive broadcast receivers usually have small metal cans on their IF transformers. Radio signals being what they are, they propagate through space and sometimes the only reasonable way to stop them is to use some kind of conductive material as an enclosure.

Superheterodyne receivers can suffer interference if signals on the IF are picked up directly instead of going through the conversion stages. Many parts of the transceiver need to be protected against electrostatic or electromagnetic components of radio-frequency signals generated both internally and externally.

Inadvertent electrostatic coupling

Electrical capacitance would, at times, be a lot more convenient if it could exist wholly in capacitors and leave other circuits alone. Small amounts of capacitance sometimes allow the transfer of enough RF energy to degrade circuit performance or even cause malfunctions. Amplifiers acting as oscillators and filters with poor stop-band performance are common examples.

The top of figure 16-1 is a simplified example of capacitive coupling. Two conductive plates supported above the ground plane form an air-insulated capacitor. A complete ac current path exists from the generator, through the plates, through the detector, and through the generator back to the ground plane.

In real circuits, the generator may consist of all manner of circuits which develop an RF potential. As for the detector, it may or may not be a circuit designed to work with RF signals. Audio and dc circuits using devices with semiconductor junctions can malfunction when invaded by an unwanted RF signal.

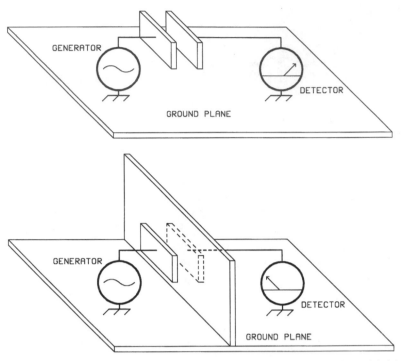

■ **16-1** *The transfer of energy is stopped by an intervening shield wall.*

Assuming the coupling is unwanted, the bottom arrangement in figure 16-1 illustrates how it can be greatly reduced. If a shield wall is placed between the two plates, the detector's plate is not only hidden visually, it cannot electrically "see" the generator's plate either. Most of the RF current now flows from the generator to its plate, along the shield and over the back-plane, returning to the generator.

Usually the shield and ground plane are made of highly conductive material. Almost no voltage exists on the shield because a very low impedance exists between its surface and the ground plane. Consequently, very little RF potential exists on the detector's plate.

At least a couple of circuits in the prototype transceiver use a simple shield similar to that in figure 16-1. Crystal filters used in the second IF are mounted in small metal boxes. Another larger shield separates the entire second IF board from the BFO/detector board. FET inputs on the second IF board are some of the most sensitive to stray capacitive/electrostatic coupling because of high impedances.

Figure 16-2 is a view of the prototype upside down, resting on its top with the bottom cover removed. That which resembles spaghetti leftovers is actually a number of insulated hookup wires lying against the motherboard's bottom layer.

Built on perforated board stock, the interface and control board shown in figure 5-24 is mounted on the motherboard in the upper-right corner. This side of the motherboard is not copper clad and makes a convenient place to glue tie-points made of small squares of unetched circut-board stock.

Two hinges support the front edge of the motherboard and the rear edge is supported by brackets at two corners. This allows the motherboard to swing upward, as in figure 16-3. Below, another thinner enclosure houses the three-chip computer and synthesizer. The transceiver is also resting on its top in this photo.

This is one example in which the enclosure is used to keep signals generated by the computer and synthesizer away from receiver circuits. The power amplifier and output filters have their own enclosure mounted on the back.

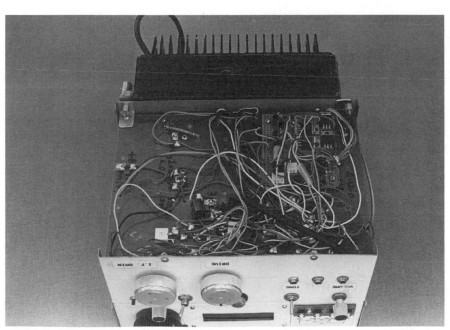

■ **16-2** *This is a view of the unclad side of the motherboard. Various tie-points and components are glued to this side with cyanoacrylate adhesive.*

■ **16-3** *Hinges made of number-12 copper wire allow the board to swing upward. A ballpoint pen is used here as a prop. More elaborate hinges would allow more movement.*

Inadvertent electromagnetic coupling

Another form of accidental coupling between circuits can, at times, be even more of a headache when circuits designed to handle widely ranging power levels at high frequencies are in close physical proximity.

Circuits which carry large currents generate strong electromagnetic fields. Sensitive, low-power, low-impedance circuits tend to be the most susceptible when located nearby. Some method of canceling the electromagnetic field is often necessary to keep the low-power circuits from being affected.

Figure 16-4 is a simplified example of how one coil connected to an ac source can induce current in another one. The top half of the conductive enclosure bordered by lines A through F could be placed over the lower half.

If only one edge—say, A—makes electrical contact with the lower part of the enclosure, the top cover would serve as an excellent electrostatic shield. Unfortunately, it may not perform well as an electromagnetic shield.

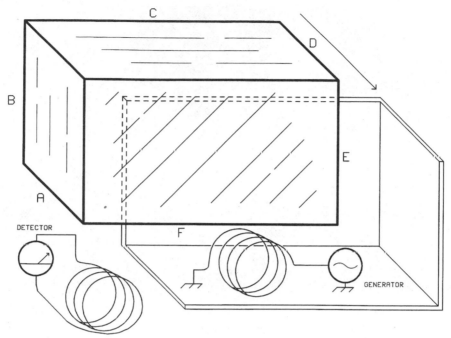

■ 16-4 *To effectively cancel electromagnetic fields, currents in the shield walls must be free to circulate. Split or incomplete shielding will not allow this if the gap is at right angles to the coil turns or parallel to the magnetic field lines.*

Nonferrous materials such as copper, aluminum, and brass allow magnetic lines of force to pass through them. Even thin sheet steel allows significant magnetic coupling when used as described.

A most effective method of reducing power transfer between the two coils is to make sure that edges A through F are electrically connected to the lower enclosure. The connections should have low resistance and run the entire length of each edge.

A rapidly changing field from the generator's coil sets up currents circulating around the enclosure interior. These currents develop a magnetic field, which tends to cancel the one coming from the coil. As a result, very little energy is transferred to the detector's coil.

As long as currents can circulate to set up opposing fields, some compromises in the enclosure can be allowed. Holes have little effect if they are insignificant in size compared to the signal's wavelength.

Figure 16-5 is a more direct view of the motherboard. At bottom left is a small enclosure made of sheet brass with soldered seams.

■ 16-5 *This is a view looking down on the baby boards. Starting at the left, the following boards are mounted running front to back: HF converter, 67- and 77-MHz oscillator, 72-MHz IF, 4.2-MHz IF, shield wall, BFO and detector, audio board. The small board at the lower right, running left to right, is the transmit preamp board. Directly behind it is the CW generator.*

It covers the second local oscillator board, which corresponds to the schematic in figure 4-10.

Relatively strong fields are emitted by tuned circuits in the oscillators and other parts of this board. Because of the frequency and power level involved, this turned out to be one of the more difficult circuits to isolate from the others.

All sides of the board are enclosed. The copper ground plane of the motherboard forms the bottom. Flanges on the bottom edge of each wall have machine-thread nuts soldered to allow bolting the box and motherboard together.

Figure 16-6 shows how three aluminum chassis and a bottom plate enclose various sections of a home-built receiver. Top, middle, and bottom sections correspond roughly to sections housing synthesizer, IF/audio, and RF circuits. Overall shielding from outside electrical fields is very good at HF.

With the aluminum enclosure, good electrical contact is maintained along each edge because of numerous screws. The proto-

■ 16-6 *This receiver, built on three aluminum chassis, is well shielded. It is also the one mentioned earlier, incorporating a digital frequency counter connected in a feedback loop with the VFO.*

type transceiver uses a painted steel enclosure, which does not make good electrical contact along all edges.

If you are already a seasoned experimenter, you may be thinking, "Gee, this transceiver could really use better shielding." That's certainly correct. Although, in many respects, its performance is equal to or better than that of a number of commercially manufactured radios, more and better shielding of most circuits would be a good idea. To gain insight into how to better shield and isolate various circuits, we need to cover a few more topics.

Traveling wave and waveguide

Relatively large signals, greater than 30 mW, at VHF frequencies are present in circuits concerned with the first and second local oscillators. In addition to the fundamental frequencies, a significant amount of harmonic energy is generated.

When RF energy is generated at a wavelength comparable to dimensions inside the enclosure, the enclosure may act as a waveguide. When this happens, fields inside the enclosure can be much stronger in some places than if no enclosure is used. As a practical consequence, birdies (false heterodynes) not heard when the transceiver's receive circuits were first tested spread out over the workbench later appeared when it was mounted in the enclosure.

Real waveguides—at least those intentionally designed to carry RF energy—are designed to convey RF energy as an electromagnetic wave so that wires and insulators with their attendant losses need not be used. The more compartmentalized your transceiver, the less trouble you will have with unwanted RF traveling in this manner.

Capacitors are often used on power, control, and audio leads to bypass RF energy to ground. When a wide spectrum of RF energy must be bypassed, it's best to use two or more capacitors with different values. This is because capacitors large enough for bypassing low frequencies usually have too much parasitic inductance at higher frequencies. Also the smaller-value capacitors have too much reactance at lower frequencies.

If RF signals are allowed to flow unimpeded along the wrong leads, the effect of good shielding will be defeated. Frequencies which would have been greatly attenuated by walls or enclosures, will be reradiated by wires or paths entering an enclosure.

When maximum attenuation is needed under difficult circumstances, feed-through capacitors are often used to route conductors in and out of enclosures. Feed-through capacitors utilize three terminals: one on each side of the shield wall and the body, which is soldered or fastened to the shield.

In some units, the capacitance is formed by coaxial conductors with an insulating dielectric in between. The outer conductor connects to the shield wall and the inner conductor extends to terminals at each end.

Other feed-through capacitors use two metal disks separated by a dielectric. One disk connects to the two end terminals. The other disk has a center hole which allows one of the end terminals to pass through. It connects to shield wall.

On the left of figure 16-7 is a sectionalized view of how a feed-through capacitor might be fabricated and mounted. The disk seen on the right makes a very-low-impedance connection to the shield wall.

Instead of a solder connection, some feed-through capacitors have a threaded body which allows fastening them to the wall with a matching nut. Threaded units are easier to use with aluminum walls.

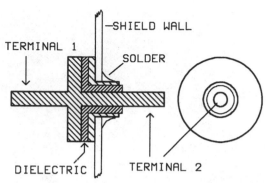

■ **16-7** *The dielectric thickness of this hypothetical feed-through capacitor is exaggerated. Also, a protective layer of insulation may be applied over the ungrounded plate.*

Figure 16-8 is a photo showing the ends of several feed-through capacitors mounted on the same receiver as in figure 16-6. Small RF chokes are also connected to provide added attenuation. I highly recommend using this type of bypassing construction for your transceiver project.

Packaging

Using many small circuit boards to build the transceiver gives you great flexibility in choosing how to design and arrange physical characteristics of the transceiver. You may want to make the transceiver lower and wider or taller and narrower. Various proportions could be realized as long as they are within reason.

I decided to mount the small RF, IF, and audio boards edgewise against the motherboard to allow simultaneous access to both sides of the boards. This worked fairly well for making measurements and adjustments, although it was somewhat awkward because of the closeness between boards.

Modifications are another matter. Changes near the top edge could sometimes be accomplished, but working near the bottom edge close to the motherboard necessitated detaching the smaller board.

Another possibility exists for those who plan to etch their own circuit boards. Arrange the board patterns in a suitable fashion and copy them all at once to make one or more large patterns. If you do this, leave plenty of room between patterns to mount shield walls should you decide to use them.

■ **16-8** *A number of feed-through capacitors are used here to pass control signals through an aluminum bulkhead.*

However you decide to arrange boards, try to keep gain blocks of the same frequency in line so that you don't have an output doubling back near an input. Circuits designed to use widely different frequencies are less likely to suffer from accidental mutual coupling.

If you don't mind the time, expense, and size, each board or section could be housed in its own small aluminum box, complete with connectors and any necessary controls. Although it may look a bit strange, this method can result in a well-shielded but flexible project.

Speaking of connectors, I used very few connectors to interconnect radio circuits. The synthesizer/computer compartment contains numerous connectors. In my experience, connectors seem to be less reliable in circuits experiencing very low voltages. Examples are found at the input of a receiver RF amplifier, IF amplifier, or audio amplifier.

When developing the synthesizer and computer interface, I often found it necessary to temporarily disconnect one section or another. Logic and control levels of several volts are less susceptible to connection vagaries than microvolt-level signals.

Another construction consideration to keep in mind concerns electrical bonding to structural parts. Dissimilar metals often come into contact because making everything out of the same material is not always practical or perhaps possible.

Aluminum enclosures are popular, but aluminum screws are not at all common. Structural connections possessing good electrical conductivity may change over time. The time span can be disappointingly short when highly dissimilar metals are in contact in a humid environment.

If the structure's electrical connections may not be top quality, you should strive for a good return (ground) path independent of the main enclosure. The prototype's motherboard with a solid ground plane is an example.

For the best shielding enclosure, the ideal situation is to have all internal circuits grounded at a single point on the external enclosure. This is true even if the outer enclosure is an excellent conductor. Unfortunately, this is not usually possible because of necessary external connections.

If all of this information about shielding and decoupling sounds intimidating, it may help to know that the prototype is mediocre in this department yet functions well. With a little care, the average builder can do much better.

Thermal considerations

While we are examining blemishes and warts in the prototype's construction, let's consider another problem: managing heat flow. Generating the largest amount of heat is the RF power amplifier, followed closely by the driver stage. These circuits are mounted against the large heat sink at the back.

The heat-sink fins run vertically and do a good job of dumping heat into the air through convection cooling. If you prefer, the final/driver stages and heat sink could be mounted on top of the transceiver. Bottom-mounting would be a poor choice, as convection currents could rise to "cook" other transceiver circuits.

Components inside the transceiver also generate noticeable amounts of heat. Transistors driving the VCO tuning motor are mounted directly above the VCO clock oscillator. Circuits below the computer/synthesizer compartment are also generating heat. These factors contribute to frequency drift.

Once the operating temperature stabilizes, the indicated frequency is stable to within 50 to 100 Hz, as long as temperatures in the shack are reasonably stable. Warm-up drift can be as much as

500 to 600 Hz. While this is not a disastrous amount of drift, especially after warm-up, it certainly could be improved.

If you use a custom-ground crystal, carefully temperature-compensate the CPU oscillator and do a better job of isolating the oscillator from heat sources; a total frequency drift of no more than 10 to 20 Hz should be possible under normal operating conditions. This will require a CPU oscillator stability of around 1 or 2 Hz because of the approximate times-10 multiplication effect of the phase-locked loop.

Access and maintenance

A big motivation for building projects like this must be curiosity, along with the desire to learn new ideas and apply them. If that is true, the project may never really be finished. A change here, a new idea there, or an improved part is often added to projects even after they've been "finished" for some time.

Even before starting on the first circuit board, you may want to try to assimilate as much information as possible, record measurements of planned enclosures, and make preliminary sketches. Allow enough room for small unforeseen changes or errors, and especially for access, when making measurements.

Projects such as this, which are repeatedly disassembled and assembled, need fasteners made for the task. Many ready-made enclosures come with self-tapping screws, which cut coarse threads into the enclosure the first time they are used.

Self-tapping screws are fine for small, inexpensive projects where the need to access internal assemblies is infrequent. I don't recommend them for a large involved project. The female threads soon wear out, especially in aluminum enclosures. You can then use a larger-size screw, but larger screws soon do the same thing and this process cannot be repeated indefinitely.

With steel enclosures and thicker gauges of aluminum, suitable machine threads can be cut into the material with a handheld tap and wrench. This procedure requires a little practice. If you've never tried it before, get some instruction, wear eye protection, and practice on scrap material first.

Installing female threaded fasteners on thin-sheet brass, copper, and copper-clad circuit boards is easy. Just drill a hole, temporarily mount a machine nut with a matching screw, and solder the nut in place. Be careful not to get solder on the screw. Remove the screw when it cools.

You can even solder nuts to steel and aluminum enclosures, although the process is usually a little more involved. With aluminum, a special flux is needed because common resin fluxes do not remove the very thin but tenacious layer of aluminum oxide which prevents bonding. Unless you are soldering a very small piece, the amount of heat needed may seem tremendous compared to most electronics soldering operations. A large, thick piece will probably require a torch or a several-hundred-watt soldering iron.

You can probably solder some thin-sheet steel pieces with ordinary solder, flux, and tools if the surface is very clean. Although the usual 60/40 electronics solder forms good electrical contacts, it is mechanically not very strong. Shops specializing in welding supplies often carry various solders and fluxes which are stronger and suitable for bonding various metals.

The motherboard, which holds most of the radio-circuit "baby" boards, is fairly convenient for allowing easy access. If the motherboard assembly is an example of a more accessible area, the motor-tuned VCO is not. You might want to refer to figure 11-4 for a visual review of the motor-driven VCO assembly.

After removing the top cover, getting to the VCO board requires several steps. First, the whole assembly has to be lifted out of its nest. The motor's spring-arm nut must be loosened and the drive wheel removed. Two small machine screws holding the variable capacitor to the case must be removed.

Finally, four machine screws, which fasten the board and metal box together, must be removed. Fortunately, the VCO box doesn't house a lot of circuitry and requires little disassembly.

Human engineering

It is entirely possible to build a radio with excellent technical specifications which is also a real pain in the neck to operate. As an example, many military HF radios have well-designed circuits, but some have no convenient method of continuous tuning, as would a an amateur transceiver or even a normal broadcast receiver.

Changing frequencies in large or small steps is relatively easy with this design. Hopefully, you will find the design convenient with no serious inherent shortcomings. If you do, change it. That's part of the fun of having your own unique creation. Letting others know about your innovation is also very helpful.

One possible way to inconvenience the operator with this project is to place controls and indicators in awkward positions. Another is to use knobs or switches which are difficult to manipulate. Figure 16-9, a front view of the prototype, is a useful view for considering where controls should be placed.

At the upper right on the front panel is a large, dark-colored knob attached to the optical rotary shaft encoder. A better position would have been below the center of the display module or perhaps even lower.

The four function keys on the left will be covered by hinged strips of metal or plastic to give the feel and appearance of piano-key buttons, when I finally get to the task. There is really no need to require the operator to push tiny buttons on a desktop-sized radio.

A hinged bail on the bottom allows for elevating the front, as in the photo, or lowering it. With an unlit LCD, the viewing angle for best

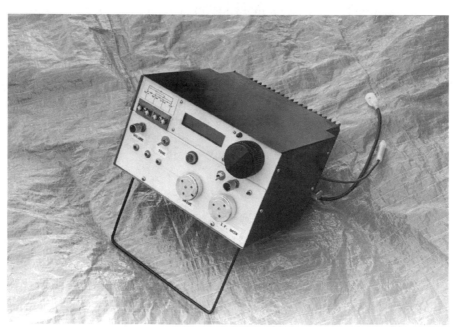

■ **16-9** *A frontal view of the transceiver prototype with all covers in place. The large knob to the right is the main tuning control. Top and center is the display module opening. Four function buttons connected to the microcontroller are at top left. The small knob directly below the function buttons is for audio gain. Three jacks at lower left, proceeding left to right, are mike audio, transmit/receive, and audio out. The large knob at lower center is for transmit drive and the similar one at lower right is for IF gain. The small knob at far right controls the SSB transmit limiter. The nearby toggle switch selects IF filters, and the other left-mounted toggle switch controls the audio response.*

contrast is more critical. Backlit LCDs are more attractive and easier to read at different angles. Even with an illuminated display, using an adjustable bail or front feet might be a good idea.

Except for the shortcomings mentioned above, I don't have too many regrets about how the transceiver turned out. Rather than use more complicated construction, I decided to accept a few compromises and complete the project.

Fabrication

Once you've decided how the enclosure and supporting structure should look, you face the task of actually making it. Actually, most hobbyists probably don't make the decision in such a straightforward manner.

For myself and others, the decision is more of a back-and-forth operation in which we consider which tools, materials, and processes are available. Next, we try to decide if they will accomplish the desired end result with a reasonable amount of effort and expense. If not, then it's back to the first step, where we rethink the enclosure's design.

Some builders produce beautiful enclosures from scratch by cutting and bending every piece of metal to closely measured dimensions. Most of us pragmatically pick out suitable ready-made boxes, then measure, mark, and cut the necessary holes.

Measuring, marking, and other shop practices are another subject area that could occupy an entire book. A little research in the library or at the bookstore wouldn't hurt if you feel the need for more knowledge in this area.

Most texts concerning sheet-metal work will describe the best or most professional methods of performing certain operations. Unfortunately, some methods of fabricating sheet metal require expensive equipment.

In order to keep down cost, you may be interested in a few cheap tricks or shortcuts for working with sheet metal. Large to medium-size round holes seem to be a problem for builders. I'm referring to holes of ⅜-in-diameter and larger.

Drilling holes in thin metal with common twist drill bits, which cut a conical impression, often causes problems. Unless the workpiece and bit are held in completely rigid alignment, the resulting hole is usually irregular instead of round. A ridge is usually left, which must be deburred. These bits also tend to bite too far into the metal and grab, which either stalls the bit or mangles the workpiece.

Socket punches are the accepted method of cutting clean holes on small to medium-size jobs. They are also a bit expensive for most hobbyists. One method of drilling larger holes in thin aluminum is to use a sharp woodcutting bit. Hole cutters such as those used for mounting door knobs can be used for really large holes.

Use a much lower speed than would normally be used for cutting wood. Always back up the aluminum with a smooth wooden block and, if possible, use a drill press. The workpiece should be clamped securely in place. If the bit or cutting teeth should seize, an unclamped workpiece flailing at your hands with its metal edges is a likely result.

Using woodcutting bits on steel is really pushing this idea a little too far. They don't do a very good job, overheat, and very quickly become dull. Although it's not as fast, a nibbling tool is an economical method of making various-sized holes in aluminum and steel. Nibbling tools are also great for cutting out rectangles and other odd shapes.

To finish out many jobs, a small assortment of files is almost mandatory. A multipurpose twist-style bit, which uses a small, angled tip extending past a larger, flat cutting surface, is becoming popular in many stores. These bits, similar to a classic pilot bit, are a good choice for drilling many kinds of thin material.

I have one more idea that may be useful if you are trying to put a painted finish on an aluminum surface. The usual approach is to sand the surface, degrease it, and spray on zinc-chromate primer and subsequent coats of enamel.

Even with thorough surface preparation, good painting technique, and oven baking, builders are sometimes disappointed with resistance to chipping. In addition to mechanically roughening the surface, you can try chemical etching with leftover circuit-board etchant.

I've tried using ferric chloride and it does seem to make a difference. I make no claims to know what the optimum etching times, solution strength, or temperature are. I know that when used full strength, it's a fairly violent reaction, so you will probably want to use a diluted solution. Also, the etching action will probably not be uniform unless the surface is first sanded and cleaned.

VHF, UHF, and microwave applications

Although the transceiver project is designed for an upper limit of 30 MHz, it has attributes which may interest experimenters working at higher frequencies. Developing some of these uses would be a trivial undertaking, while others involve major hardware and software modifications.

The easiest way to use the transceiver at higher frequencies is to connect it to an external transverter. A more complicated topic involves modifying the synthesizer to operate at higher frequencies for use in other systems.

Using the transceiver with transverters

Many transverters require an input level of about 1 mW, and figure 5 1 at the C1-L1 junction would be a good place to tap in. For reception, connect to the D11-D12 junction. Remember to provide a way to disable the output amplifier.

Of course, if you don't need the HF capabilities, you could dispense with the time and expense needed to build the transmit preamplifier, driver, final, output filters, and TR switch. This is assuming your transverter uses a 28-MHz IF because of the low-pass filter in figure 5-1. Receiving capabilities for frequencies below 30 MHz are retained.

Higher-frequency PLLs

The following involves some advice for those who may want to adapt the synthesizer for use in systems at VHF and higher. The synthesizer can be used directly at 144 MHz with almost no hardware modifications. Probably all you will have to do is reduce the inductance of L1 in figure 11-1.

The software is, of course, a different matter. Twenty-two bits in FMML are used to hold frequency information in the present program. The upper 2 bits, which specify the emission mode, could be freed if you don't need to store the mode in memory. A 24-bit value would allow using the present display resolution at 144 MHz.

If your system uses an IF offset, the constants between NOTM and DOOFF should be changed. Actually, if all you want to do is just get the synthesizer running, great heaping portions of code could be discarded.

Going to higher frequencies requires ever more software modifications to accommodate the larger numbers. For those with experience in designing for these higher frequencies, the hardware could be modified to operate at frequencies as high as 2 GHz with essentially the present topology.

An MC12031B could be used as the prescaler. The fine-tune loop would need a tuning range for the VXO of only 81 Hz or so of range for the same top reference frequency of 16,384 Hz. This would require an impracticably long time in the COUNT routine to measure the synthesizer mixer output with enough precision to use 10-Hz resolution.

A much better solution would be to use 50 or 100 Hz as the smallest tuning step and use a much higher reference frequency. A higher reference frequency is easy to implement by software and would improve the PLL's performance, making it less susceptible to transients, microphonics, and close-in phase noise.

For narrow tuning ranges such as a single amateur band, you may want to dispense with the motor-actuated coarse-tuning capacitor. On the other hand, oscillators designed to cover wide frequency spans might benefit from motor actuation of a sliding tuned cavity resonator. A surplus oscillator using an yttrium-iron garnet (YIG) resonator might be a more compact solution.

Different intermediate frequencies

Using different intermediate frequencies for HF operation should be one of the more easily accomplished design modifications. The first IF could be moved up or down a few megahertz with no software changes other than putting in new offset values.

More drastic changes, such as using a high IF of 9 MHz, would require redesigning practically the entire transceiver and software package. Ready-made crystal filters in the 8- to 9-MHz range are popular in commercial and home-built equipment. Using crystals in this range for the lower IF is entirely reasonable and would not be a very difficult design modification.

By carefully choosing the second local oscillator frequencies, you may be able to pull off this change without a software change. The prototype switches from lower to upper sideband by changing the second local oscillator frequency. This is because the second IF crystal filter has an asymmetric passband shape and is used as a lower-sideband filter only.

Most commercially made crystal filters have a symmetrical passband shape and could be used for upper-sideband operation by switching the BFO frequency. With upper- and lower-sideband generation, only one second local oscillator would be needed.

Although the control board would continue to receive different data for USB, LSB, and CW modes with unmodified software, signals from the control board could be rerouted to switch BFO frequencies instead of second local oscillator frequencies. Slight modifications to the control board might also be necessary; however, if you are more into hardware than software experiments, this may be relatively easy.

Other modifications and ideas

Although almost anything can be changed when building your own gear, two topics are relatively simple. Changing the tune-rate selection involves nothing more than substituting some different constants.

Another small bit of code is an experiment which allowed me to control frequency memory selection by manually turning the coarse-tune VCO capacitor. Finally, some ideas about adding FM are discussed.

Change tune-rate selection

As mentioned before, the tuning rates in the Main Program Listing (see appendix A) are 10 Hz, 100 Hz, 10 kHz, and 1 MHz. These choices may not suit everyone. Starting around line 1227 at NOTRC is where you can find constants moved into TMP_1, TMP_2, and TMP_3.

If, for instance, you want to try 1-kHz increments instead of 10 kHz for the f (fast) tuning rate, no problem. Change the code in the top part of the following to the bottom example:

```
NOMED: CJNE A,#00100000B,NOFAST
       MOV TMP_1,#0E8H
       MOV TMP_2,#03H
       MOV TMP_3,#0
       MOV CURC,#01100110B ; "f"
; Change code to this.
NOMED: CJNE A,#00100000B,NOFAST
       MOV TMP_1,#064H
       MOV TMP_2,#0
       MOV TMP_3,#0
       MOV CURC,#01100110B ; "f"
```

The hex values are 10 times less than the actual increment because the values are later multiplied times 10 to obtain the actual frequency values.

Manual VCO range control

At line 594, labeled NOTRD, is an interesting bit of code which is preceded by semicolons. The semicolons cause the assembler to treat the lines as comments and not assemble them as instructions. If the semicolons are removed, allowing the instructions to be assembled, a new 3-byte frequency memory selection is made each time through the loop if the PLL is not locked.

When the VCO's main tuning capacitor knob is turned to a position which puts the VCO within lock range of a frequency memory value, the locked condition can be detected by P1.1. In figure 10-4, R21 and C21 are to be connected to P1.1 for this purpose. If P1.1 is high, searching stops.

This modification in effect allows you to use a knob on the VCO coarse-tune capacitor as a memory selection device and eliminate the need for a motor drive. This technique may be worth looking into if you are interested in building a synthesizer which will consume as little space and power as possible.

How well does it work? If the frequency values are spread well apart, such as one entry per amateur band, it's not too bad. When two frequencies are within lock range of each other, picking out one can be difficult. Perhaps code could be added which would provide temporal separation between entries when close frequency values are scanned.

FM mode

Some builders may have an interest in adding frequency modulation (FM) capability to the transceiver. FM is used in the higher-frequency portion of 10 m and it could be handy to have when using VHF and UHF transverters for local communications.

You can easily change the mode indication from AM to FM by changing line 1379 following LSBNO in the Main Program Listing (see appendix A). The present binary value of 01000001 is the ASCII value for a capital "A." The ASCII value for a capital "F" is 01000110 binary or 46 hexadecimal.

That is the easy part. Generating FM is slightly more work and receiving FM may be the most challenging part of adding this mode. Frequency-modulating an RF oscillator signal is relatively easy.

The same sort of circuits used to remotely tune the VCO in a phase-locked loop can be used. It's basically a matter of providing a bit of speech audio to a varactor diode. The main problem usually is maintaining good oscillator stability while achieving the necessary deviation and linearity needed.

Very stable oscillators using quartz crystals are not easily pulled off frequency. If the amplitude of speech audio is increased to make the oscillator deviate more from its undisturbed frequency, keeping a linear frequency-to-voltage relationship becomes more difficult and a distorted-sounding signal may result.

Usually when crystal oscillators are used for FM, the oscillator frequency is multiplied a number of times in frequency multiplier stages before being amplified and used at the output frequency. The transceiver uses a mixing scheme for frequency conversion, which means that using a crystal oscillator at the first IF is out of the question. Modulating one of the second local oscillators is a possibility.

Perhaps the best place to introduce FM is at the synthesizer output VCO. You could reroute audio from the existing speech amplifier to the tune line going to the varactor diode in the VCO.

A larger capacitor would have to be temporarily connected to the loop filter to slow its response during FM transmissions. If this isn't done, the loop will quickly correct deviations, nullifying the speech audio. A 0.1-μF capacitor should be a good starting value. The larger capacitor should be disconnected during operation in other modes.

As for the control board in figure 5-24, it should be modified to turn on the lower IF carrier generator. The audio stages are already set to mute when transmitting. Assuming you intend to replace AM with FM capability, perhaps the circuit involving Q8 can be modified to drive circuits which would perform the two switching functions mentioned above.

Adding FM reception will require even more work. Although FM transmitting capability can probably be added with several simple modifications, adding receiver circuits will require whole new subsections or circuit boards.

The easiest approach to adding the FM receiver circuits is probably to put one of the receiver-on-a-chip systems, such as the Motorola 13135/136P, on its own small board. Use the 72-MHz signal at R8 in figure 4-5 as the input point. The second local oscillator should be turned off in receive mode.

Audio from the added board can be routed to the transceiver's audio board, bypassing all of the SSB/CW IF circuits. I hope these ideas are of assistance to those of you determined to build FM capability into your transceivers.

A spectrum analyzer project and test equipment issues

WOULDN'T IT BE NICE IF WE NEVER NEEDED TEST EQUIP-
ment? (Unless you manufacture test equipment, I suppose.) What
if the quest for reliability in electronics reaches perfection?
Except for accidents or other external damage, electronic equipment would never break or malfunction. There would be no need
to track down an elusive problem.

When building, we would all create impeccable new designs, consistently taking into consideration every nuance of nature. Possessing an exhaustive knowledge of all applicable physical laws
and making no mistakes would eliminate any need to examine the
circuit with instruments to see if our plan has flaws.

Returning to reality, we really do need test equipment when things
go awry or when we need to measure some electrical parameter.
Sometimes it is easier or more practical to measure and adjust circuit values than build in the correct value to start with. An example would be narrowband tuned circuits in IF strips.

Certainly not all test and adjustment procedures require test gear.
Often, behavior of the device under test is an indicator. Tuned circuits are often adjusted by listening to sound levels or watching
the meter indication on a radio or other device.

Test gear does not seem to be the staple commodity in ham
shacks that it once was. When radios glowed in the dark, almost
everyone had a volt-ohm-milliampmeter (VOM) and many phone
operators had some kind of oscilloscope. With today's reliable,
integrated solid-state equipment, some operators don't even
bother to purchase an inexpensive multimeter to make simple
measurements.

I assume because you are reading this book, you're not a 100 percent appliance operator or at least don't intend to stay that way. It is very unlikely that anyone can successfully build a complicated radio transceiver without some test equipment. The more skillful you are at using test equipment and interpreting the results, the more likely your success.

Many points in the transceiver and other circuits exhibit waveforms with complex shapes. Any shape other than a sine (or cosine) wave means more than one frequency is a component of that waveform. Measuring the amplitude or even viewing the waveform on an oscilloscope may not give you sufficient information in some cases.

It is often necessary to know which discrete frequencies are combining to make up a complex waveform and what their individual levels are. Measurements in the frequency domain performed with a spectrum analyzer may look something like figure 17-1. The face of a cathode-ray tube (CRT) has images drawn on it by manipulating the electron beam so that horizontal movement corresponds to frequency changes and vertical movement represents amplitude.

Other instruments can sometimes provide similar information, although their capabilities for doing so are more limited. Useful devices include wavemeters, dipmeters, and communication receivers. The following spectrum analyzer project is not very sophisticated, but its capabilities allow making some measurements much more easily and accurately than some of the substitute methods listed above.

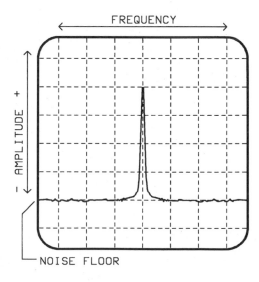

■ **17-1** *A typical example of a single-frequency signal displayed on a spectrum analyzer might look like this with vertical divisions representing 10 dB and the horizontal divisions representing 100 kHz.*

A basic spectrum analyzer scheme

Spectrum analyzers can be designed in a number of ways. One straightforward method would be to build a series of narrow-bandwidth receivers to cover the frequency spectrum to be measured. The output of each receiver would be assigned a specific horizontal position on the display. Although a scheme using many individually tuned receivers might respond quickly, it would be horrendously expensive.

Most RF spectrum analyzers are an electronically tuned super-heterodyne receiver coupled to a visual display. Figure 17-2 illustrates how a ramp generator, voltage-tunable receiver, and oscilloscope are combined to make a spectrum analyzer. The oscilloscope in this case is a stand-alone unit. Many commercial analyzers use a built-in CRT and associated circuits.

For the ramp generator, I used a potentiometer with a large calibrated dial and knob. Spectrum analyzers usually have an automatic method of generating a sweep and tuning voltage which increases linearly with time. Chapter 18 discusses a simple ramp-generating circuit in another instrument, which could be used here.

Many low-bandwidth, inexpensive oscilloscopes could be used here. The scope must also have a horizontal input capable of static

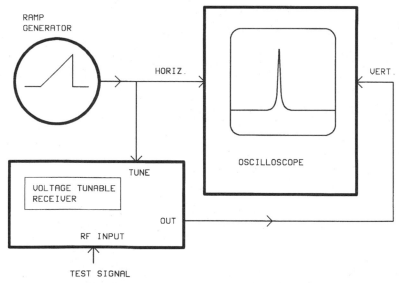

■ **17-2** *The project in this chapter uses a simple voltage-tunable receiver. Many commercial systems use more complex tunable receivers.*

deflection with dc potentials. Oscilloscopes are usually vertically calibrated in some fraction or multiple of a volt-per-division. Horizontal divisions on the graticule indicate time periods.

To interpret readings of a system like this, you will have to make a conversion table by taking measurements of known frequencies and power levels. Alternatively, you can switch from the device under test to a calibrated signal generator, duplicate the position and height of any peak in question, and read from the generator's frequency and output-level displays.

A common use for spectrum analyzers is to examine the output or interstage signal of a transmitter. Have you ever hooked a transmitter or transceiver to an antenna system and noticed that you could not get the standing wave ratio (SWR) reading down to acceptable levels, even though the very same system shows a good match when used with another transmitter?

Strange readings as above usually indicate a dirty signal, one with spurious frequencies in addition to the desired one. Although the signal may sound fine on the fundamental frequency, the selectivity of the antenna system causes reflections of the spurious component as indicated by SWR meter readings. Even more exasperating is the fact that the problem may go away when a dummy load is substituted for the antenna.

I often use a matching network similar to that in figure 17-3 with my wire antennas. It is a link-coupled network. I've experienced problems with some transmitter and transceiver projects where SWR readings would abruptly jump from relatively low values to much higher values as slight adjustments at C1 or C2 were made. The cause of these sudden jumps is, in most cases, a parasitic oscillation.

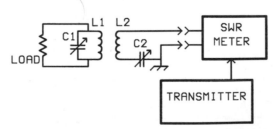

■ **17-3** *Even with the matching network adjusted for a perfect match at the operating frequency, a transmitter with a spectrally impure signal may result in reflected power readings.*

This sort of thing can be difficult to observe with some instruments. Communication receivers sometimes don't work well close in to a transmitter, even with sufficient attenuation in the input line, because of inadequate shielding. Wavemeters have a rather limited dynamic range and cannot distinguish between closely spaced spectral components.

A parasitic oscillation in the transmitter's output stage is a likely cause when the symptoms are sensitive to the antenna system. Oscillation could take place over a range of many megahertz and may change frequency as tuning conditions change.

Finding the signal or signals can be difficult with the average receiver because of slow tuning rates and too much selectivity. The usual signal strength indicator, an S-meter is also a slowly responding indicator.

Building your own spectrum analyzer

The spectrum analyzer presented in this chapter is somewhat crude, but it is useful when looking into situations such as the above-mentioned scenario. In its present form, it is not well suited to examining the close-in spectrum around a signal. Some ideas for examining narrow frequency ranges will be presented later.

Figure 17-4 shows the major building blocks of the system. Using an inductive pickup loop at the input allows you to vary input lev-

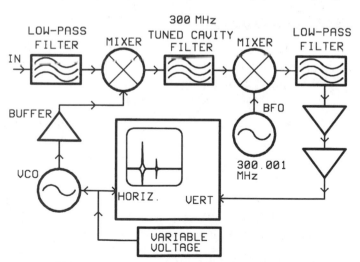

■ **17-4** *This is an overall block diagram of the spectrum analyzer project. The IF and filter operate at approximately 300 MHz.*

els. For more versatility, a variable attenuator should be used when direct connections must be made to the equipment under test.

This is actually an up-converting superheterodyne receiver with a first IF of 300 MHz. I chose a 3-dB cutoff frequency of 100 MHz for the low-pass input filter. The IF filter is just a large-diameter, air-insulated, quarter-wavelength transmission line.

Two frequency conversions are performed. Normally, the second IF might operate at several tens of megahertz so that a bandpass filter following the second mixer could select either the sum or difference signal in the mixer output and reject the other as an image. Otherwise, two responses would be generated for each spectral component of the input signal.

The philosophy behind this design is: "What the hey, let it generate two responses and put them so close together that they look like one pip." Another simplification is that I didn't include a detector or use logarithmic amplifiers. That's why the responses in figure 17-4 go above and below the line in the oscilloscope block.

Linear or logarithmic amplifiers and detector circuits could be built with op-amps because the second IF needs to respond to signals only as high as the widest resolution bandwidth—probably 100 kHz or so. Resolution bandwidth refers to the minimum frequency separation between two signals of equal levels so that there is a 3-dB dip in the display curve between the two signals.

Converter construction

After any necessary attenuation is performed by an inductive pickup or an in-line resistive attenuator, the input signal in figure 17-5 passes through a low-pass filter and then to a 3-dB attenuator before reaching pin 1 of the SBL-1 mixer.

The local oscillator at Q1 feeds the input of an MWA0304 amplifier chip whose output is applied to pin 8 of the mixer. The mixer output passes through another resistive attenuator before it reaches the IF bandpass filter.

All of the RF circuitry for this project is mounted on a solid, unetched ground plane. Figure 17-6 shows the pin-out information you will need to wire the HF converter.

The SBL-1 drawing represents pin numbering as seen from the bottom of the device. Leads on the amplifier chip and transistor at

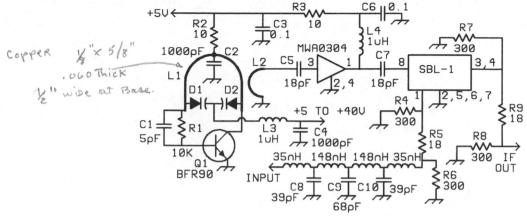

Copper ⅛" × 5/8"
.060 Thick
½" wide at Base.

■ 17-5 *These components convert HF signals to 300 MHz. D1 and D2 are SK3327. Lower-capacitance diodes should be used for situations requiring lower tuning voltages. This may be necessary if you use the ramp generator circuit discussed in chapter 18.*

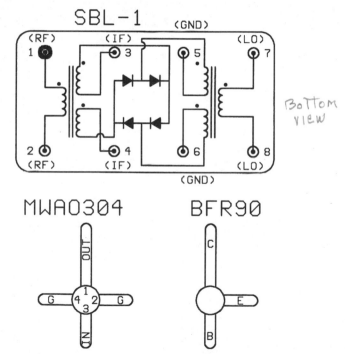

■ 17-6 *Pin 1 of the mixer has a blue dot around it. This is the numbering sequence as viewed from the bottom. The transistor and IC are keyed to the longest lead.*

the bottom can be identified from either side because each device has one long lead.

If you have never built a project using a solid ground plane with no etched paths, this is probably not a good candidate for your first attempt. Some people also refer to it as dead-bug or ugly-board construction. Because of the frequencies involved, some leads must be very short and direct. Nevertheless, I will try to describe how some of the components should be mounted.

Components likely to cause the complications in positioning are those with more than two leads. You may want to make a pencil sketch of the intended layout. Another possible method is to gather all the components and make a mock-up by stabbing their leads into a styrofoam block. You can then use a sketch of the mock-up or simply remove parts one at a time as you remount it in a similar position on the real board.

A good place to start might be Q1. It has only one grounded lead, the emitter. I mounted Q1 so that the flat sides of the transistor body were vertical with the emitter lead pointing straight down. About ⅛ in or less away from the transistor body, the emitter lead should be cut off or bent at right angles. Of course, if you are making a mock-up, just stab the emitter lead into the styrofoam block.

The collector and base leads are horizontal, parallel with the plane of the board. If you want, 100-kΩ (or higher) resistors can be soldered from base to ground plane and collector to ground plane to act as supports.

C1 and R1 can be mounted next or at least measured to find out how much space they will need to connect to D1 and L1. L1 is a small U-shaped piece of sheet copper. A heavy-gauge wire could be substituted. L1 is a ⅛- by ⅝-in strap of copper 0.060 in thick. The ends are separated by a distance of approximately ½ in.

C2 supports the middle of L1. The junction of D1, D2, and L3 is so rigid after being soldered together that it should not need extra support. Once you get the hang of it, this kind of construction is simple and fast. Solder grounded circuit components to the board and use the other end as a support where possible. If you run into a situation where support must be provided, use a high-value resistor.

Leads 2 and 4 of the MWA0304 should connect directly to the ground plane. With the IC body lying flat against the board, take care not to short the input or output leads to ground. A mica insulator under the leads could be used to guard against shorting.

The mixer is mounted on its side. I suppose you could mount it on its back with the leads pointing upward; however, it is probably easier to arrange the layout for short, direct leads by using side-mounting.

Presently, the oscillator operates from below 300 MHz to 375 MHz. This means the spectrum analyzer has a useful input range of below 1 MHz to over 75 MHz. Apparently, response to the lowest input frequencies is diminished because energy from the VCO buffer rides through the IF filter and overloads the mixer or IF amplifiers. This is also the condition where the VCO is tuned very close to the 300-MHz fixed local oscillator, within the passband of the IF filter.

Link inductor L2 is nothing more than the body of C5, its leads, and the ground plane. The solder connections, as with the other RF components, is very close to or flush against the body, with practically no lead length.

You can vary the orientation and spacing between L1 and L2 to affect power transfer from the oscillator. Coupling should be no greater than the amount required to produce about 75 percent of the maximum output level available from the MWA0304.

The VCO tuning voltage is specified as +5 to +40 V instead of 0 to +40 V, so the tuning diodes will not be forward-biased at the low end of the voltage scale. You can use a voltage source which starts at 0 V as long as a 2- to 10-kΩ resistor is included in tuning line. Oscillation will probably cease when the diodes are forward-biased, but this is not a problem because the frequency will be below 300 MHz at that point.

The oscillator and buffer should be well shielded with power and control leads bypassed to prevent its signal from being picked up by other parts of the analyzer. It is also a good idea to shield the mixer output lead and components from the input lead. Miniature 50-Ω coaxial cable is used to connect to the IF filter.

IF and second conversion

As you can see in figure 17-7, processing of the 300-MHz IF signal is accomplished by a bandpass filter, mixer, and oscillator. The output of this oscillator is fed directly to the mixer. Another buffer amplifier may be useful here.

I'm really not sure how much the mixer can pull or destabilize the oscillator. A free-running LC oscillator at 300 MHz is rather sensitive to load and supply-voltage changes.

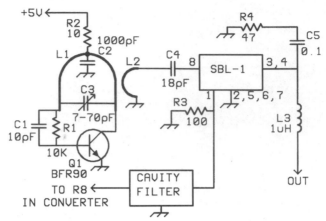

■ 17-7 *After passing through the cavity filter, the 300-MHz IF signal is converted to a low frequency by the SBL-1 mixer.*

Most of the oscillator circuit is like the VCO except C3 replaces the tuning diodes. Gain variations in transistors may necessitate trying different values at either R1 location if an oscillator doesn't behave.

A single resonator made of 1.5-in copper pipe filters IF signals. Construction of the filter depicted in figure 17-8 is straightforward but requires tools and techniques more common to plumbing than electronics projects.

The best way to cut the ½- and 1½-in copper pipe to length is with a tubing or pipe cutter. If done properly, this will leave smooth, square ends. Two ¾-in-diameter copper disk form capacitor plates at the free end of the ½-in copper pipe. A threaded brass rod or machine screw is used to adjust the capacitance. The nuts should be brass or copper.

A stationary disk is soldered to the small pipe. The movable disk is soldered to a nut, which is also soldered to the brass screw. A stationary nut is soldered to the end plate. A free nut on the other side is used to lock the movable capacitor plate into position.

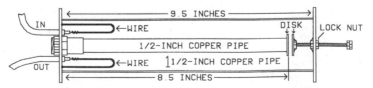

■ 17-8 *In the 300-MHz IF filter, you can use number-12 or -14 wire for the coupling loops soldered to the input and output cables.*

I used a ½-in sweat-to-NPT adapter and a ½-in conduit nut to support the small pipe. Sweated plumbing connections are soldered by putting together the pipe and connector after they have been cleaned and flux has been applied. The joint is then heated while applying solder to the connection and filling the space between each piece. Other types of connections, including soldered connections, could be used to fix the pipe to the end plate.

If you use thick copper stock for the end plates, the large and small pipes can be soldered to them. I didn't do this for several reasons. Doing so takes a lot of heat—and solder. Soldering may also take a lot of time unless you use pipe end caps instead of flat plates. Using so much heat would probably loosen solder connections on some of the smaller parts. With everything soldered together, it would be difficult to disassemble.

Instead, I used sheet copper backed up by plywood blocks. The blocks are compressed against the pipe ends by nuts on the ends of four 10-24 threaded rods which run the length of the assembly. The rods are evenly spaced around the circumference of the pipe. The electrical conduit nut is also tightened against the block.

Soldering to the thin sheet copper is relatively easy and the filter is easy to disassemble for maintenance or modifications. The coupling loops are 2 in long, and I really don't know if they are optimum for this application.

To be honest, the whole project is one interesting experiment. Although it provides useful measurements, many areas could be improved. Some weak false responses can be seen on the display.

Higher-performance mixers with more carefully matched terminations may help. Energy from the 300-MHz fixed oscillator causes weak false responses in the first mixer because it is not completely balanced out. An IF amplifier stage with high output-to-input isolation, such as the Mini-Circuits MAN-1AD, could be used to reduce this backward signal flow.

Too much gain at the 300-MHz IF would not be good because it would cause more opportunity for intermodulation distortion. The extra gain could be canceled by a resistive attenuator in the amplifier's output path with a beneficial increase in isolation.

Baseband amplifier

Following the second mixer, filtering, and amplification in figure 17-9, process the signal so that it can be viewed on an oscilloscope. The far right op-amp must be a wide-bandwidth type if you plan to

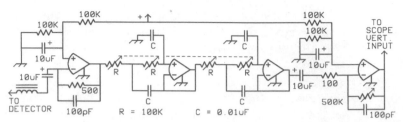

■ **17-9** *Pin numbers are not specified because they will vary with the type of op-amp.*

have full gain at the highest setting of the 500-kΩ potentiometer. A cascade of op-amps with a lower cutoff frequency could be used if a smaller-gain pot is connected between the output and inverting input.

By using manufacturers' data in selecting op-amps, you can select those with a suitable gain-vs.-frequency curve. Simply make sure the gain, which you control with feedback resistors and capacitors, is set lower than the curve for open-loop gain at the frequency at which it will be used.

As gain is increased by changing external feedback, the maximum frequency reached before output drops off is lowered. Between this point and some low frequency, gain will be constant. Gain at the lowest frequencies is reduced because of reactance in the coupling capacitors.

The actual amplifier circuit presently in use is not the same as in figure 17-9. The present amplifier is connected to LC filters using components which are not readily available. Figure 17-9 is a suggested circuit, which should make up for some of the present deficiencies such as limited bandwidth selection.

If you build a system like this, some sort of calibration should be provided for R. Dual-ganged pots are readily available and various means can be used to gang the two units together. An alternative to mechanically connecting two pots is to use a bank of resistors with a four-pole rotary switch.

Although the amplifier and filter circuits are slightly different, everything else in figure 17-10 is as described here. Space at the back of an aluminum chassis holds the 300-MHz IF filter. Power supply components are located at the left.

Two enclosures made of sheet copper are bolted to the top of the chassis. The smaller one toward the front panel houses VCO and mixer circuitry. Another mixer and fixed oscillator are mounted

inside the larger copper enclosure. Amplifier and filter circuits are mounted on small boards at the extreme right.

If you use a manually operated pot for controlling the VCO, a calibrated dial for the knob is handy. Even if you include a ramp-generator circuit, the pot should be included so that you can switch to it for static positioning of the display trace and the frequency. This will allow reading the fundamental frequency of an incoming signal with a frequency counter.

Using other instruments

An oscilloscope is used with this project and the swept-frequency generator in the next chapter. The oscilloscope is a valuable and versatile instrument in its own right. It can display voltage relative to time. Most oscilloscopes use a cathode-ray tube display although some have become available with other types of displays, such as the LCD.

A small area appearing as a dot on the display indicates the relative amplitudes of voltages which are used to move this dot. Although horizontal movement can be effected in other ways, the most common method is to use an internal circuit to generate a voltage which increases linearly with time.

The dot is normally moved from left to right. When it is time for a new trace to begin, the horizontal deflection voltage quickly drops to its beginning value, placing the dot at the left. During this action of snapping back to the left, the trace is not visible.

In cases in which other displays are used, pixels are addressed by manipulating the horizontal position in a similar manner instead of using analog voltages. In either case, vertical positioning results from the voltage being measured.

Many oscilloscopes use purely analog means to manipulate the positioning of a dot formed on the face of a CRT. Linear amplifiers build the signal to a voltage level sufficient to drive the deflection plates of the CRT. A focused electron beam from the hot cathode passes between these plates on its way to the phosphorescent and fluorescent inner surface of the CRT face.

Other types of oscilloscopes convert input signals at the vertical amplifier to digital values. Once the value is converted to a binary number, it can be manipulated in many ways and/or used to address pixels on a liquid-crystal display.

One important reason for converting the input signal to a series of digital values is to put them into memory where they can be stored indefinitely. This is important when a very fast event occurs infrequently or perhaps only once.

A pattern on the CRT must be energized long enough for the observer to get a clear view. When the electron beam is moving very fast, this can be accomplished by retracing repeatedly over the same area. Most scopes have the ability to trigger the start of the horizontal sweep from an external signal. This allows synchronizing the horizontal sweep with a repetitive waveform.

Various methods of holding the image on the CRT face have been used with analog oscilloscopes. A long persistence phosphor continues to glow long after being energized by the electron beam.

Nowadays, the most common scope for capturing fast, infrequent events is a digital storage oscilloscope. It can "replay" a single event over and over as if it were a repetitive signal. Many can interface with a computer or printer to store files or print out waveforms.

Another popular way to obtain a digital storage oscilloscope (DSO) is to purchase an adapter/interface board for a personal computer. If you already have a computer, some of the oscilloscope functions such as display and controls already present in the computer hardware help provide DSO capabilities at a lower price than a comparable stand-alone unit.

Software for some "computer" scopes can even perform calculations on values of the waveform samples to give a frequency-vs.-amplitude display, thus acting as a spectrum analyzer. Discrete points along the input waveform are sampled by digital scopes with an analog-to-digital conversion performed at each point.

A simple superheterodyne receiver with a low IF could be combined with a computer-based spectrum analyzer to obtain a very narrow resolution bandwidth for examining HF signals. By using the computer-based spectrum analyzer on the receiver IF, the system could be much less expensive than a full-featured stand-alone analyzer with similar resolution bandwidth specifications.

The better digital scopes have higher sampling rates and higher resolution when making the conversions. If the sampling rate is not high enough, very fast changes between samples will not appear. Specifications for digital oscilloscopes differ in a number of ways from analog scopes and you may want to learn more about them. For instance, the sampling rate must be much higher than the input frequency and is not equivalent to the bandwidth rating of an analog scope.

Instrument checks

A wavemeter is an instrument which is capable of providing coarse but useful readings of frequency and relative amplitude. The frequency-determining part consists of a tunable resonator such as an LC-tuned circuit. The resonator is connected to a detector and microammeter or other indicator. Even a small incandescent bulb can be used as an indicator for higher power levels.

Inductive or capacitive coupling brings RF energy from the circuit under test to the wavemeter. Another instrument known as a dip-meter or dip-oscillator can be used to detect when a circuit is resonant at the oscillator frequency.

Some dipmeters can also be used as wavemeters when in a nonoscillating condition. Dipmeters and wavemeters are useful for coarse measurements. They are also often useful for checking the validity of measurements made by other instruments.

Digital frequency counters are a convenient and accurate instrument for measuring frequency. Unfortunately, they sometimes display erroneous values because the input waveform is distorted. Many general-purpose counters have a triggering level adjustment to help pick the correct point on the waveform for incrementing the counter.

Various circumstances sometimes make it difficult to know which value is correct when you can get several different readings by adjusting the triggering level. Usually, the correct reading is the lowest frequency displayed. The situation may be even more confusing if you are using a small handheld counter with no triggering control. Other readings are most often harmonics or multiples of the lowest reading, but not always.

Situations can arise in which nonharmonically related frequencies appear. This might be the case with the output and even the input terminals of some mixer circuits. Connecting the counter to the circuit under test may load it and change its operation. Reflections can occur in the counter input cable, or high input levels may overload the counter's input circuits and distort the signal.

Such conditions may accentuate frequencies other than the one you are trying to measure. You may be working on your own project and have documentation or an understanding of what frequency to expect. If so, at least you will know when the numbers look reasonable.

When you have difficulty obtaining a correct reading, you can try placing a resistor at the probe end of the counter's shielded input cable. This resistor, in combination with the cable capacitance, acts as a low-pass filter. This often reduces high-frequency components of the signal enough for correct triggering. The resistor also reduces loading of the circuit under test.

The actual resistance value depends on many factors, such as the counter's sensitivity, input impedance, and input-cable capacitance. With a sensitive counter, 1 kΩ is often a reasonable starting point. A circuit under test with high output levels will probably require a higher resistor value than a lower-power circuit.

In some cases, it is helpful to use another instrument and gain more information about what is really being measured. If you are fortunate enough to have a high-frequency scope or spectrum analyzer that covers the necessary frequency range, either one can be used to measure the frequency closely enough for a check against the counter.

Of course, developing projects at home often means working with what you have. Fortunately, frequency counters capable of gigahertz operation are now reasonably priced. Oscilloscopes capable of operation at VHF are pricey. Most spectrum analyzers are also rather expensive for a hobbyist's budget.

A wavemeter or similar instrument may provide the sanity check needed for frequency counter measurements without breaking the budget. Another type of wavemeter which may be more appropriate at VHF and higher is made of parallel-conductor transmission line. Wavemeters made in this manner are known as *lecher wires*.

Figure 17-11 shows how two bare copper wires are mounted ½ in apart on wood. Machine screws and nuts hold tension on the wires in a manner similar to that of a stringed musical instrument. The wires form a half loop at the other end of the wooden beam where they are anchored.

The wires in this setup are 4 ft long, but they can be any length you wish. It all depends on the wavelength you will be testing for. Power is coupled into the line when an inductor carrying RF energy is placed near the shorted end of the line.

With the line open at one end, a standing wave pattern can be observed by measuring RF voltages between conductors at different points along the line. This is assuming the frequency is high

■ **17-11** *Both ends of the lecher wires are supported by ¼-in wooden blocks mounted on the wood beam.*

enough to allow voltage and current nodes to develop. In other words, the line must be longer than a half wavelength.

Instead of moving an RF voltmeter along the line, it is often more convenient to vary the electrical length of the line while observing what effect this has on the circuit to which it is coupled. You can vary the electrical length by sliding a shorting bar along the line. It should be insulated from your hand and held at right angles to the line.

This is how I used the lecher wires in figure 17-11 to check oscillators in the spectrum analyzer project. When monitoring the dc current supplied to the oscillator, it would peak noticeably as the shorting bar traveled between points separated by half-wavelength distances.

If you are measuring the half-wavelength distance in inches, the frequency in megahertz is:

$$\text{Frequency} = \frac{5905}{\text{distance in inches}} \qquad (17\text{-}1)$$

Measuring in meters:

$$\text{Frequency} = \frac{300}{\text{distance in meters}} \qquad (17\text{-}1)$$

A swept-frequency generator for crystal-filter evaluation

SOME RADIO-BUILDING TASKS INVOLVE ADJUSTING INDUCtive or capacitive reactance for a desired response. Peaking simple tuned circuits is a common adjustment. Many older radio receivers have front panel controls for the preselector tuned circuit, which has a single point where received signals sound loudest.

Adjusting this type of filter, which has a single, relatively broad response, is simple. It is easy to tune by ear, listening to a steady signal. Even multielement IF filters, such as those used for narrow CW bandwidths, often turn out well with little difficulty.

Making good SSB filters from mass-produced crystals can be more of a challenge. Computer software is available to help design the filter. Most catalogs don't list the crystal unloaded Q or equivalent series resistance which you will need to design the filter. In addition to testing the overall filter, the system presented here can be used with a frequency counter to check these parameters.

The crystal filters in this book were designed and tested using the swept-frequency generator shown in figure 18-1. It was also used to check the response of the filter after being installed. This allowed me to make sure it was terminated correctly and performing as expected.

Getting help

Much of the fun in building projects lies in the fact that you are doing many things for the first time. With so much new and unexplored territory, it's easy to run into a situation where progress is stymied by one problem or another. The impediment could be in

■ **18-1** *The swept-frequency generator and amplifier/detector system are in the foreground. A general-purpose power supply hooked to the system sits on the right. A large knob at the front of the aluminum minibox is used for coarse frequency adjustment. The tuning line on top of the box is connected with an alligator clip to allow other kinds of frequency-modulation experiments.*

the planning stage or it could be trouble in acquiring parts. Often, troubleshooting is required to bring a balky project to life.

Many amateur experimenters enjoy answering questions of a technical nature. We get to help someone else while sort of participating vicariously in another project. Ask around; flatter someone by asking for help if you get stuck. If you don't know a nearby technically competent amateur, you might try contacting the American Radio Relay League (ARRL) to see if one of their volunteers, known as a Technical Advisor, could be of assistance.

Obviously, asking for help should be done in a responsible way. Learn all you can about the problem before getting other busy people involved. It's not very nice to bring others an electronic disaster because you haven't learned how to solder. Schedule things at their convenience and don't expect them to practically build the project for you.

The following project first appeared in the March 1994 issue of *QEX*. A revised version is printed here with permission of the American Radio Relay League.

Why build it?

What motivated me to construct this test instrument? Did the effort exceed the rewards? Not when I build or otherwise deal with projects using crystal filters.

Now that I have an easy and comprehensive way of viewing the results, the construction and alignment of crystal filters is easier and less intimidating. The system described here takes much of the guesswork out of applying salvage or new crystal filters. It adds to my success in quickly building filters from inexpensive microprocessor crystals. Honestly, it's not that hard to build either.

An overview

The basic setup is shown in figure 18-2. The voltage controlled oscillator (VCO) is primarily a conventional LC-tuned Hartley oscillator. In operation, the frequency is tuned over a small range by a varactor diode. Figure 18-3 shows that I used an MV2104 type. Substituting other types is certainly permissible as long as the capacitance specifications aren't too different. If you don't get the sweep width you want, try changing the 5-pF coupling capacitor to compensate.

After going through a buffer amplifier, the VCO signal is sent to the filter under test. From there it goes to a wide-bandwidth amplifier and on to the detector. Having been rectified and filtered, this varying dc voltage drives the vertical input of an oscilloscope. At any particular point in time, the deflection and sweep circuitry commands the VCO to "run at this frequency." Meanwhile, this same

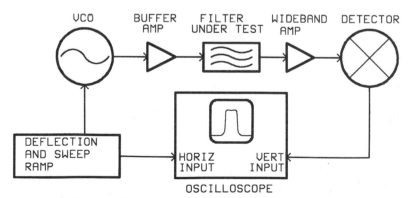

■ **18-2** *All of the important ideas are here. If you have or find the right subassemblies, you could reduce the amount of component-level construction needed to finish the system.*

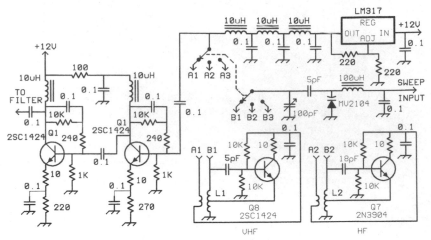

■ 18-3 *I placed all of this circuitry (plus two more oscillators) in an aluminum box. Everything else goes into an identical box bolted to the first. Signals before the filter are kept separate from those after the filter, eliminating any leakage. If the VCO signal is allowed to "leak" past the filter under test, the stop-band attenuation will appear worse than it really is.*

deflection voltage is causing the oscilloscope beam to deflect left or right to a position corresponding to the aforementioned frequency.

You don't have to build a self-contained system as I did. A commercial sweep generator would eliminate everything but the wideband amplifier and detector. Unfortunately, many older analog units don't have sufficient stability and sweep-width resolution. Newer synthesized generators do, but reasonably priced ones have an upper frequency limit that may be too low for many filters. One option is to modify a conventional analog frequency generator. The necessary sweep range is small enough to make this a relatively easy task.

Of course, assembled wideband amplifiers and detectors are readily available. Motorola, Mini-Circuits, and many others have suitable devices. If you really want to simplify the horizontal sweep generator, just eliminate it and leave the "tune pot." This is possible because the sweep-rate repetition must be rather slow—about 1 Hz or less. If the frequency is swept too quickly, the passband curve displayed by the oscilloscope becomes too distorted. The problem now becomes one of finding a person to crank the pot back and forth while you work on the filters. Good luck!

My generator/detector system originally covered a range of frequencies of approximately 6 to 74 MHz in three different ranges. It uses a separate RF oscillator module for each tuning range (figure 18-4). Figures 18-3 and 18-5 show three sharing a common variable capacitor for the coarse tuning adjustment. The VCO output and supply are multiplexed on a common line and switched by one pole of a multiple-position rotary switch.

Two of the oscillator coils are wound on PVC plastic pipe. The other is self-supporting and made of number-14 copper wire for the highest frequency range. Although PVC forms with Super Glue for dope may not be state-of-the-art technology, frequency stability is completely adequate for this instrument.

Physical and electrical dimensions are as follows:

The large coil has a 0.85-in inside diameter with a length of 1.1 in. It has 18 turns of AWG number-28 wire and the inductance is 5.32 μH.

The inside diameter of the medium coil is 0.85 in. The length is 0.55 in and is covered by seven turns of AWG number-22 wire, resulting in a 1.35-μH inductance.

■ **18-4** *This is a close-up view of one of the RF-oscillator modules that forms most of the VCO. I used ground-plane construction for all of the modules. Support points for some of the components are made of tiny bits of PC board glued or soldered to the main board. The large dark piece of enameled wire is the output link for the oscillator.*

■ 18-5 *This is the compartment housing the VCO and buffer-amplifier circuits. It's a 4- by 2⅛- by 1⅝-in aluminum box from Radio Shack. VCO modules are on the left. The coarse-tuning capacitor and varactor diode are in the upper right. Buffer amplifiers along with the band switch and decoupling network are at the lower right. The buffer-amplifier board is also of ground-plane or ugly construction. The VCO output links are wired with miniature 50-Ω teflon coaxial cable.*

The small coil measures 0.5 in at the inside diameter with a length of 0.75 in. An inductance 0.27 μH is achieved by winding five turns of AWG number-14 wire.

The two larger coils are wound on ¾-in CPVC pipe stock. The small coil is wound by using a ½-in drill-bit shank and removing the bit.

Coil turns are not evenly spaced on the forms because final tuning (to obtain overlapping frequency coverage) is accomplished by compressing or spreading turns. This should be done with a frequency counter connected to the buffer-amplifier output before cementing the turns in place.

Although this arrangement gives continuous coverage from 6.9 to 74 MHz, other builders could end up with slightly different results because of stray capacitance in the circuit. Various contributory sources include: circuit layout, minimum capacitance of the main tuning capacitor, and the range switch.

Three oscillator modules cover the frequency range mentioned above. Both ends of the frequency range can be extended by adding more oscillator modules. The high-end module should be located as close as possible to the switch and variable capacitor.

You may even want to put a fixed capacitor in series with the B lead of the oscillator where it connects with the main tuning capacitor. Oscillators sometimes jump into a different mode of oscillation if component lead inductance is a significant portion of the resonator inductance. Reducing total capacitance in the resonator will allow using a larger coil if this becomes a problem.

The lower end of the frequency range is easy to extend. It is also the mostly likely to need added coverage because many crystals and filters exist below 7 MHz. Although it is not shown in the photographs, another oscillator module was added, extending coverage down to 4 MHz.

The added module uses the same circuit as the HF module in figure 18-3 except the tapped inductance is larger. Actually, the inductor is physically smaller but has a larger inductance because I used a small powdered-iron toroid.

Lower-frequency oscillator modules tend to have a higher harmonic content. This is not usually a problem with crystal filters unless you mistakenly test a filter at the wrong frequency. Adding a resistor between the emitter and center tap of the inductor reduces overall gain and reduces the harmonic content.

The oscillator and buffer stage are operated at low power levels to minimize frequency drift caused by heating of circuit components. A buffer amplifier is definitely a necessity. Crystal filters cause large load changes as the frequency is swept in and out of the passband. These large changes in impedance tend to "pull" the oscillator frequency and cause inaccuracies in the passband shape depicted by the oscilloscope.

Much of the buffering action is brought about by adjusting the link on each oscillator module for light coupling. The less power taken from the oscillator, the less effect load changes will have on the frequency.

The buffer amplifier should be stable and have a relatively constant amount of gain across the frequency range in use. That is why feedback networks exist in the amplifiers. Although this scheme does a suitable job, other amplifier circuits might be more appropriate for the buffer. Two or three stages with interstage load

285

resistors and frequency-compensating networks should have higher output-to-input isolation.

The wideband amplifier in figure 18-6 is more or less lifted from pages of the ARRL *Solid State Design for the Radio Amateur* by Hayward and Demaw (see Bibliography). Of course, you can now do the same job with integrated circuits and save quite bit of time (see Ward in Bibliography).

The detector uses some forward bias for D2. I first tried a simple unbiased diode detector but found the dynamic range limited to about 50 dB. Adding some dc bias increased the dynamic range to almost 70 dB. At that point in development, the detected RF signal was simply fed to the vertical input of the oscilloscope. Unfortunately, with a mostly linear scale on the vertical display, the higher attenuation levels (larger negative dB numbers) become crowded together at the bottom. A logarithmic dc amplifier would be a good way to alleviate this problem. Not having the good fortune of such a device within the confines of my junk box, I found a different solution. Diode D3 across the detector output (scope input) allows you to increase the vertical amplifier sensitivity

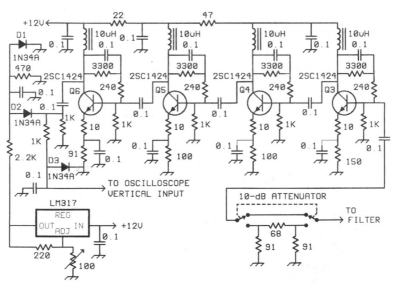

■ **18-6** *I used 2SC1424 transistors for Q3 through Q6. Other similar devices, such as the 2N2857, may be easier to find. Less expensive and more robust transistors can be used if the amplifier bandwidth will be more limited than needed here. Try to use a transistor gain-bandwidth product greater than or equal to 10 times the highest frequency to be amplified.*

while compressing or limiting the response to higher-level signals. With this arrangement, high levels of attenuation (low-level signals) are easier to observe and low attenuation levels are still visible on the CRT face.

The horizontal-deflection sweep circuit uses a dual op-amp IC. See figure 18-7. One section is used as an oscillator. The other is an integrator. The integrator output changes linearly with time, helping to give a more uniform brightness level as the trace is moved from side to side. The component values shown result in a sweep rate that is about as fast as possible without severe distortion of the passband curve. You may want to slow it down even more for frequencies above 30 MHz. Increasing C1 decreases the sweep rate. Increasing C2 decreases the slope of the output waveform ramp. A small piece of perforated board seemed to be the easiest material to use for making this part of the system, and the results are visible in figure 18-8.

Operation

The CRT is swept in both directions, left to right and right to left. The event displayed is a result of changes in frequency, not time. Therefore, I found it unnecessary to incorporate the usual right-to-left, snapback, and retrace blanking used in oscilloscopes.

The switch in figure 18-7 disables the automatic sweep function when opened. By monitoring the voltage-controlled oscillator with

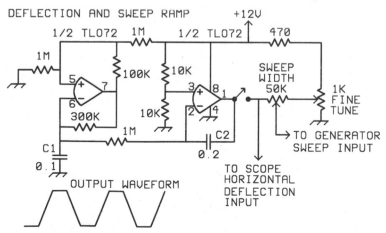

■ **18-7** *Opening the switch allows manual control of the deflection and sweep voltage. Positioning the sweep and deflection in a static manner allows you to read the VCO frequency with a counter.*

■ **18-8** *Not the neatest layout ever, but it works. Actually any style of construction is suitable for the deflection and sweep circuit as long as you make the correct connections. Although not shown here, the wide-bandwidth amplifier and detector are in this same enclosure. I used ground-plane construction techniques again. The physical layout looks somewhat similar to the schematic in figure 18-6.*

a frequency counter, you can make precise bandwidth measurements. Turn the fine-tune control to position the CRT beam at points on the passband curve that you want to measure. The difference in frequency readings is the bandwidth at that particular level of attenuation.

Oh yes, the scope face graticule isn't calibrated in decibels, so how do you find those –6- or –60-dB points or whatever? One way is to substitute a calibrated attenuator for the filter under test. Another method is to use an external signal generator at the passband frequency. Plug it directly into the wideband amplifier input. Adjust the external generator for the output level you choose as the zero-dB (no-attenuation) point. Step the generator levels down from that point, making note of where various attenuation levels appear on the graticule. If you pick the right sensitivity setting for the vertical-deflection amplifier (on the scope), 0 dB will be near the top of the graticule, with –60 dB near the bottom. I use this system on an old Heathkit model IO-4541 with the vertical deflection set for 0.16-V full-scale deflection.

The buffer amplifier is set up to drive a 50-Ω load and the wide-bandwidth amplifier input impedance is about 50 Ω. That's great if the filter is designed to use a 50-Ω termination. Of course, most are not. Various methods can be used to accommodate these differences. Wide-bandwidth transformers are one-way; lumped-constant LC networks are another, more flexible way. When I'm trying to find the best termination value for a new filter, a simple resistor or pot is usually the best way to nail things down. If the optimum terminating resistance is going to be somewhere near 50 Ω, the method in figure 18-9 works okay.

Resistors R1 and R2 add to the impedance present at the generator output and wideband amplifier input. If, for example, the filter needs an 80-Ω termination on both ends, R1 and R2 should each be equal to 30 Ω. You can extend this method of termination only so far. The limit will depend on the output level of the generator and sensitivity of the wideband amplifier.

With the present setup, large resistance values (over 100 to 200 Ω) may cause too much attenuation. For such situations, see figure 18-10. Under some conditions, capacitance of the field-effect transistor can affect the test results. This happens when the capacitive reactance is low enough to significantly affect the apparent value of R2—in other words, when you are dealing with high frequencies and/or high R1/R2 values. If so, use the L1,C3 circuit. L1 and C3 are tuned to resonance at the passband frequency of the filter under test. The combined capacitive reactance of C3 and the FET gate are opposed by L1 and create a very high impedance looking into the gate.

Applications

OK, so you build this system or something similar—then what? It is valuable for tasks other than just building new receivers, transceivers, or SSB transmitters.

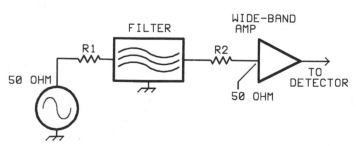

■ **18-9** *R1 = R2 = desired filter termination value – 50 Ω.*

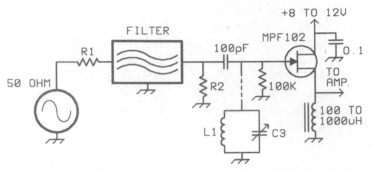

■ 18-10 *R1 = desired filter termination – 50 Ω. R2 is the desired filter termination. L1 and C3 are optional. Use them if the FET gate capacitance will lower the impedance enough to appreciably affect the filter (see text).*

Substituting crystals or entire filters different from those specified in a published design can degrade performance. I like being able to see these changes on the oscilloscope trace, as opposed to tedious and inaccurate sessions of tweaking by ear.

Maybe you're thinking of trying something new or experimental. How about using varactor diodes in a ladder filter to control the

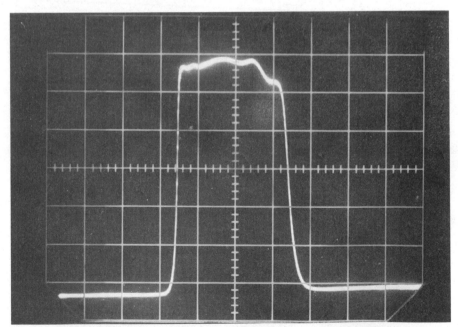

■ 18-11 *Caution—filter under construction! This photo was taken before clamping diode D3 was added. The passband is about 10.7 MHz and the horizontal divisions represent approximately 1 kHz.*

A swept-frequency generator for crystal-filter evaluation

passband shape remotely, or perhaps using overtone responses for a high intermediate-frequency filter?

So, none of this is on your things-to-do list? One popular way to improve the performance of commercial receivers and transceivers is to change or add IF filters. Filters designed by the manufacturer or aftermarket suppliers are usually plugged or soldered in and ready to use. But what if you want to try a different brand (the steal picked up at the swap meet)? Cascading filters with maybe an added stage of gain are another possibility if room is available. In rigs that don't have one, a tail-ending filter after the IF amplifier is a good way to reduce wideband audio noise. All of these things are easier to accomplish with this system. You can easily see the effects of filter termination changes, tuning adjustments, etc. Passband ripple is easy to observe. You're not forced to painstakingly tune back and forth while watching a meter face and frequency readout. There's a lot of information in figure 18-11. Get it the easy way; make beautiful (?) pictures with your scope.

Scoping out a scope

The vertical and horizontal signals from this system drive the corresponding inputs of an oscilloscope. A low-bandwidth (cheap) scope is suitable if the horizontal sweep (not sync) input is dc-coupled. If you are looking for an inexpensive scope, don't overlook this point. Many are ac-coupled. They are fine for many purposes but will not allow the very slow and static horizontal displacement you need for use with this system. A quick flea-market-style test goes like this. Set the horizontal-sweep deflection to external input. Connect a small battery to the horizontal-input terminals. The spot of light on the screen should deflect immediately left or right. With dc-coupling, the spot stays there as long as the battery is connected. An ac-coupled input will allow the spot to slowly drift back to the original position. Some oscilloscopes are specified as having X-Y inputs. This means one of the amplifiers normally providing vertical deflection can also be used as a horizontal deflection amplifier.

Power distribution and operating techniques

ALTHOUGH SOME OF THE TRANSCEIVER HARDWARE IS designed for SSB operation, it also does well when used for CW. Various methods of voice communication including SSB are certainly popular now. CW transmissions have much to offer that may not be appreciated by someone unfamiliar with the mode.

CW has an inherent advantage over SSB signals under similar circumstances. Because the bandwidth for CW transmissions is much smaller, the signal-to-noise ratio is higher.

Because a CW signal packs all of its power, including keying sidebands, into a width of 100 Hz or so (this depends on keying speed and rise time), less noise power occupies the same frequency range.

An SSB signal would need approximately 2.7 kHz. Assuming both signals are exposed to the same random noise, 27 times as much noise power rides in with the SSB signal.

Another phenomenon which makes the CW advantage even greater is a type of filtering or signal processing which is accomplished by the ears and brain. People can hear and identify single tones in noise better than complex sounds like voice.

Learning code

Unfortunately, some people get off to a bad start with learning international Morse code telegraphy and develop a firm opinion that it is difficult and not at all enjoyable. This is too bad because most of them are perfectly capable of mastering it and probably with less effort than they imagine.

I don't mean to imply that you can do so without effort. If you are not spending some time at it every day or so, that is a different matter. I'm referring to those poor souls who spend a great deal of time and effort only to find that, at some point, progress in speed and accuracy seems to stop.

The subject of practice time is an interesting one. We now have many different kinds of code practice aids available. Motivation is a critical element when it comes to getting the necessary time. I firmly believe all hams interested in learning code should attempt to do some of their practice on the air.

Computer-generated code is a valuable resource; however, it can become rather tedious if it is the only way you listen to code. This shouldn't be so mysterious. Most of us get to talk back when we are learning a verbal language.

Amateurs of any license class can use code for two-way communication. The code-free Technician license allows all of the same privileges as those granted to higher-class license holders for the bands starting at 50 MHz and above. If you hold a code-free Technician license, you can use CW or any other legal modulation, such as tone-modulated narrowband FM, to communicate. This is assuming you are proficient enough to at least identify your station and obey FCC regulations.

No sending test is now required for any amateur license. The optimistic assumption is that those who can copy code at the required rate can also send readable code. I urge all instructors to try to include some hands-on sending instruction with a straight key for beginners.

Even better, put them on the air after they have some sending competency. If the student is not licensed for the frequency in use, he or she can act as an unlicensed participating third party under supervision of a control operator. I've seen this kind of experience generate a lot of excitement, especially when HF is used and a random contact can be far away from the local area.

Actual on-air experience is also good training for the code test. Static, fading, interference, and different sending styles can add difficulties that increase the stress level just as testing does with many students.

After they get over the initial fright of their first real on-air experience, some students begin to look at CW operations with great interest. Of course, there is no reason why other sending methods should not or could not be used. Sending with an electronic keyer or computer keyboard is fine.

If at all possible, you should record your sending occasionally. See if you can copy it well and compare it to examples of properly sent code such as W1AW transmissions or computer-generated code.

The important thing is to make yourself understood at the other end. Use whatever it takes and learn to send good code. If you can use a straight key, electronic keyer, and keyboard, so much the better. Versatility is nice but not essential.

Avoiding roadblocks

It is my opinion that when the progress of a dedicated student stalls, it is usually because the individual is trying to use the wrong learning process. Different skills, if you want to call them that, are learned in different ways.

I think it is really better to think of learning code as acquiring a set of trained reflexes. There are a number of bad—really bad—ways to learn code. They all probably require using more highly developed portions of the brain. Too many people who know that I'm a reasonably proficient CW operator will likely attest to that fact.

Memorization is not the way to learn code. You may need to remember how the character is formed during the first few minutes when you are introduced to it, but its makeup of short and long parts is best forgotten as quickly as possible. It should be remembered as a sound and possibly a feeling. Sitting down and memorizing the patterns in the whole alphabet is one of the best ways to sabotage your progress.

The code should not be learned as combinations of dits and dahs or dots and dashes. It should not be visually memorized as short and long marks on paper. Visualization is useful with code learning but not in this manner.

When code classes are taught, a lot of information may be retained in an inconvenient and awkward form. Unfortunately, the students may not realize it until they have difficulty copying at higher speeds.

This is not to say that instructors or students are remiss. It is natural to internalize information in ways which proved useful for acquiring other skills. I think instructors and students who have some insight into what should happen at the very start of instruction will be in a better position to avoid learning roadblocks.

Don't expect to use reasoning to decipher the code. Each letter has a certain sound and rhythm. Association is also a potential problem. Associating a previously learned character with another because it is similar often causes mistakes. The student may transpose the letters when speeds increase.

See what's happening

A very useful visualization technique is to see the letter in your mind when you hear the character. Different opinions exist as to how you should record information which you are copying. Some suggest printing and others say use a cursive writing style. Most of the really fast operators use a keyboard or typewriter.

However you expect to see the recorded character is probably the best way to visualize it. With more experience and as your copying speed increases, you will realize that you are also hearing and visualizing entire words. This is especially true of the shorter, commonly used words.

In copying characters, you will be using previously learned skills and associating them with the sounds of code characters. For instance, you already know how to write an "A." As you write more and more "A"s to the sound of "dit dah," an association forms between the "A" sound and all of those motor and sensory actions involved in writing an "A."

To be honest, I'm not really convinced that such associations are the best way to reach higher speeds. I suppose if you are going to make associations with anything, at least this takes you directly from a sound to a written character. It is probably going to happen anyway, at least during the early part of your learning curve.

The problem with letting your writing motions help you remember a character is that it is a character-at-a-time procedure. For higher speeds and much more enjoyable CW operation, you need to reduce your reliance on such associative actions.

No more paper?

How do we wean ourselves away from the character-at-a-time habit? That's easy—throw away the pens and pencils and sell your typewriter or keyboard. Well, at least put them aside. People who really know how to enjoy CW operating have advocated following this advice for a long time.

When you are under the pressure of copying in a classroom or in a real QSO, or are even experiencing the self-imposed pressure of better copy from a canned source (computer or tape), it is easy to want to cling to your pencil. I did. I didn't make a serious effort to copy in my head until I was forced to.

No, I didn't break both arms. I more or less fell into a trap of my own making. To provide some entertainment on a long road trip, I thought it would be fun to try operating HF mobile. At the time, the only rig that my wife Sylvia, N5IZZ, and I had which seemed suitable was a small 3-W, 40-m CW transceiver.

The mobile station was set up to allow operating the rig in the passenger position because, obviously, doing so from the driver position would be too difficult. I figured the little homebrew rig might provide some interesting listening and I might even manage to make a contact with it. I was wrong. Listening was fine, but I made dozens of contacts with the rig.

If you have traveled with a VHF rig, you probably found times and places at which it's hard to stir up activity. HF on 40-m CW is a lot of fun in a mobile. The added range made a lot more stations available.

I have a problem that indirectly helped give a big boost to my copying ability. At the risk of being a bit indelicate, it is motion sickness. Even with medicinal help, I can't do close work in a moving vehicle for too long without problems.

I found that it was not at all necessary to write everything down. Just jotting down the major items was sufficient. By the time the trip was over, I realized that most of the contacts didn't require writing any notes!

After the trip, I started operating while driving to work and on most other trips. The little 3-W rig now has traveled more than 70,000 mi while in operation most of the time.

I'm not advocating that everyone try mobile CW. Even though it has been an enjoyable and safe experience for me, it requires that you give top priority to operating the car and always make sure amateur activities never get in the way.

Paper-free copying doesn't just allow you to put words and phrases together, it almost forces you to. Driving or some other concurrent activity seems to have a way of freeing the mind from being distracted in the process of copying code. I've even heard of operators using exercise machines while operating their rigs.

Copying in your head is probably most effective for building your speed above 13 words per minute or so. At slower speeds I have trouble keeping the characters in mind long enough without being distracted. Those of you with better concentration may not forget as easily.

Even at slower speeds, you should mix some "head copy" in with your practice sessions. This fits in well with visualizing characters, and I really think it is an excellent way to increase copying speed.

When you can relax, listen to the context of on-air conversations and tinker around the shack at the same time, CW operating is much more enjoyable than being chained to the desk trying furiously to scribble each character.

If you learn code by using some of the mental props mentioned earlier, you could find a plateau beyond which progress becomes difficult. If you avoid or eliminate the props, you may still find a leveling-off point. Instead of 10 or 15 words per minute, it may well be 40 or 50 words per minute or higher.

Single-sideband operating

Most of my own experience using SSB has been by using the very transceiver you have been reading about. To be forthright, the comparative simplicity of CW equipment has been very alluring for a number of reasons. Simple CW projects were often very useful and rewarding.

Nevertheless, ambition turned into obsession and I started planning another project. SSB just had to be included in the rig. It would be silly to include all of the other features and not SSB capability.

Over the last year I've operated the rig in SSB mode much more often than CW. I still enjoy CW operating very much but felt that I needed more experience with this mode. Also lab tests don't always show problems that can happen under actual operating conditions.

SSB operations are a lot of fun. A number of practices are different. It seems to me to be somewhat of a different culture. Even the best SSB voice quality does not match that of narrowband FM communications. To Sylvia, the contrast is so obvious that she reports forming a mental picture of every SSB operator with a "cute little snout, pointy ears, and curly tail."

I feel that what we must guard against is not sounding like pigs, but acting like them. Part of the differences in customs and attitudes is probably rooted in differences of a technical nature.

A herd of SSB signals with their porcine proportions are certainly more likely to interfere with each other than narrow CW emissions. In spite of such difficulties, most operators manage to put their best foot forward without stepping in someone else's territory.

One tendency, which seems to be more prevalent on 75 and 40 m, is a strong attachment to a small band of frequencies. With some stations on 75 m, changing frequency up and down the band is a hassle. Someone using a tuned antenna system, matching device (antenna tuner), and tube-type amplifier has a major chore when making a substantial frequency change, because of all of the retuning required.

With all the work necessary to move, it's not to hard to see why a territorial attitude might develop. Perhaps technological changes will circumvent the issue. Broadband circuits and automatic tuning devices may eventually see wider use.

An unfortunate offshoot of the previous problem is an occasional tendency of a small minority to form cliques and operate in a sort of closed-shop fashion. I can think of all kinds of uncomplimentary things to say about this practice. Perhaps those who are this desperate to maintain some sort of "control" need to reevaluate their priorities.

Of course, some of my own bias is probably showing through here because I'm not really fond of roundtable operations. I prefer talking to one or two operators at a time. However, hams who frequent roundtables and nets in a cordial manner are to be admired because more people can use our bands at the same time.

My personal impression of HF operating is that it is easier to initiate random QSOs using CW than it is with SSB, given that power levels and other conditions are equal. SSB contacts are relatively easy for low- to medium-power stations whose operators enjoy group participation.

Another interesting difference between the two types of operating modes is that it seems relatively hard to find a grumpy CW operator. Perhaps the immediacy of the phone with its ability to convey tone, inflection, and mood sometimes makes it all too easy for us to get the message across.

It is slightly harder to berate someone via CW, so usually we don't bother. Occasionally, a point of irritation is simply muttered as the disgusted operator continues to send "FB ON HOUSEBREAKING THE AARDVARK, REALLY ENJOYED THE QSO, HAVE A NICE WEEKEND ES 73."

Good attitudes and operating procedures will go a long way toward making the phone bands more enjoyable. Always listen before making a transmission.

If you must tune output or antenna circuits, perform all but the very last adjustments with low power. An even better choice for tuning the antenna system is to use a receiver noise bridge which will allow matching without making any transmissions.

Make short calls, interspersed with breaks for listening. Avoid long CQs and use reasonable transmission lengths for all transmissions. Not only are overly long transmissions inconsiderate, they can be downright impractical. Propagation sometimes changes so quickly that you can end up talking to yourself without ever having a chance to find out if the other station copied your information.

Voice-operated transmit capability (VOX) can certainly help in shortening transmissions, although using a push-to-talk button on the microphone serves as well if it is used properly.

One option you may want to consider is a foot switch for putting the rig into transmit mode. Like a push-to-talk switch, it is immune to extraneous sounds which can trigger a VOX circuit. By using a boom-supported microphone, you can have both hands free.

You should speak plainly and clearly. Enunciation and diction which suffices in face-to-face conversations may not work as well over the airwaves. For spelling difficult words under unfavorable conditions, phonetics can help. In order to avoid causing additional confusion, use standard International Telecommunication Union (ITU) phonetics.

Unwanted sounds

Another problem that sometimes cuts down on communications effectiveness is extraneous noises surrounding the microphone. In some situations, such as mobile operating or crowded areas, you may simply have to deal with the noise because it can't be eliminated.

Microphones with reduced sensitivity at the sides and rear can help. Some microphones have an exaggerated response to low-frequency vocal sounds if used too close, and breath sounds can take on an explosive quality with some of them.

Whatever the microphone type, it should be used as close as possible without causing the problems described above. Wind and pop screens can sometimes help if they are not already built in. Speech amplifier gain should be set no higher than necessary.

Strive to keep the distance from your mouth to the microphone as constant as possible. This is good operating procedure under normal conditions and is even more important when your operating position is assaulted by unwanted sounds.

Operating the home-built transceiver

Some operating issues are specific to the transceiver project. Even if you have a pretty good idea how everything should work out from working with the hardware and software, you may find it helpful to compare results with the prototype.

Split-frequency operation is sometimes necessary and is selected by the appropriate push buttons, as described in chapter 13. Selecting the RIT function freezes the transmitted frequency to the value displayed when you select this mode.

The receiving frequency remains tunable to any frequency you like. That may sound extreme, but it's true. We will see what the practical limitations are shortly. XIT operation requires an additional step. When you enter XIT mode, the receiving frequency is frozen.

Turning the main tuning knob changes the transmit-frequency indication. However, the actual transmitted signal will start out on a different frequency unless you push the preview button first. This is because the synthesizer's fine-tune loop must have time to find the correct DAC value.

While the preview button is depressed, the transmit frequency becomes active, letting you hear or *preview* the frequency before you put a signal on the air. I suppose this quirk can be viewed as a feature which forces the operator to do the right thing and listen to the transmit frequency before putting out a signal.

You can use the split functions with various frequency separations, depending on the emission type and other considerations. SSB operations can be carried out with huge separations of several megahertz as long as you let the synthesizer settle on frequency before talking.

If signals are emitted too soon, they could be out of band and certainly would not be a permissible form of emission even in the band. If you plan on regularly using this capability, a lockout and delay circuit driven by the phase-lock indicator should be included.

Separations of a few kilohertz are more reasonable because the synthesizer achieves phase-lock quickly enough to prevent spurious emissions. As for CW, the prototype is set up for full break-in operation in which you can receive signals between code elements.

Separation with CW operation should be restricted to spacings which will allow the PLL to settle before the keyed carrier comes up. With component values as they are now, this seems to be about 5 kHz or less.

If you want to operate with wider frequency separations, a T/R switch can be included, as shown in figure 19-1. When the T/R switch is closed, everything is ready to transmit, much like SSB mode. No signal is emitted until the key is depressed, turning on the carrier oscillator.

With a scheme like this, any separation can be used as long as the PLL has time to settle on frequency. If you use really large spacing, as in cross-band operations, a split-second hesitation before hitting the key is necessary before the motor-actuated coarse-tuning capacitor in the PLL settles.

If while using RIT you want to swap the transmit and receive frequencies, change to XIT. You can swap back and forth between XIT and RIT as often as you wish. It is not necessary to repeat pushing the preview button unless the transmit frequency has been changed.

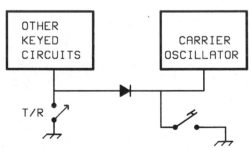

■ **19-1** *With this keying arrangement, you can use full break-in operation with the T/R switch open. This is for small frequency separations during split-frequency operation. Manually closing the T/R switch allows plenty of time for the synthesizer to settle before the key is closed for larger frequency separations.*

RF output power issues

If you are contemplating building extra circuits into the transceiver, an excellent choice would be to include an SWR/power meter for RF output measurements. I use an external meter with the prototype. It not only helps me check for a matched output impedance, it also helps when setting the drive level for transmitting.

No automatic SWR protection circuitry has been included. The amplifier has tolerated transmitting into severe mismatches at full power without damage. Antenna-system SWR and drive adjustments are most critical when using SSB at full power because you can end up with a spattering, distorted signal if operating conditions for the final amplifier stage are incorrect.

Try to operate into antenna systems with an SWR of 1.5 to 1 or less. To consistently achieve maximum output power without splattering, I set the IF clipper to maximum. This does a good job of holding voice peaks below a level which would overdrive the final amplifier.

By cranking up the drive level while continuously keying the rig in CW mode, you can see where the amplifier begins to saturate, because output stops increasing as the drive is increased. Further increases in drive level will even cause the output to decrease slightly.

Using an oscilloscope is probably one of the easiest ways to check for flat-topped waveforms indicative of distortion and splatter. I often use another crude but effective method for setting the drive level without resorting to the scope.

A loud, high-pitched whistle can be used to generate an output which simulates a CW carrier. While whistling into the microphone using a dummy load, I adjust the drive control to the point of saturation or just below it. This is as loud as any voice peak is likely to be and the clipper circuit is effective enough to assure that no speech sounds will modulate above this level.

This is not as foolproof as an automatic level control (ALC), but it provides plenty of talk power for a relatively simple circuit. Once you find the correct drive setting, you can log it for future reference. Gain of the transmit amplifier chain varies slightly from band to band, so you probably will need to record several settings for the drive control.

Arranging memories

Alphabetically tagged memories A through P allow you to put whatever frequency you prefer in each. Because they are accessed sequentially, you probably will want to arrange closely related frequencies to be close in the alphabetical order.

When power is first applied to the transceiver, the memory locations will have random values in them. The identifying character for each location may also be some random, strange-looking letter or symbol. If this happens, push the memory-up or memory-down button repeatedly until the correct sequence of uppercase symbols appears.

Once the symbols for frequency-memory locations appear to be correct, you can enter the values you want. Set the tune rate to "h" (huge) which allows tuning in 1-MHz steps to quickly cover large frequency spans. Next, punch the tune-rate button again to obtain 10-kHz tuning steps and tune in closer to the desired frequency.

Another push of the tune-rate button will activate the 100-Hz tuning steps. If you need to tune in even closer, the 10-Hz per-step increments are available with another push of the tune-rate button. Usually when you are setting up the memories, just dialing something within your amateur or broadcast band of interest is sufficient.

After you have frequencies programmed in for a number of bands, you can use memory-up or memory-down like a band switch. Using only two or three bands allows you to use several memories per band for frequently used frequencies. Instead of cranking the main tuning knob, just punch up another memory for a net or schedule.

Main tuning and the transmitted signal

While on the subject of using the main tuning knob, there is a weakness in the firmware concerning tuning rates of 10 or 100 Hz. If the knob is spun too fast, the synthesizer sometimes appears to stick at the previous frequency instead of moving.

This electronic backlash effect can probably be eliminated by future incarnations of the firmware. One simple way to reduce this tendency is to use the four-tooth wheel shown in figure 13-2 instead of the eight-tooth wheel.

In general, the transceiver is fun to operate and a lot easier to use than some of my other homebrew equipment. During the occasions when high-quality test equipment was available, out-of-band emissions were well within FCC guidelines.

One problem which turned out to be more intractable than most was related to frequency instability when transmitting. It was intermittent and caused more of a problem with SSB transmissions than CW.

The time constant in the PLL loop filter was too long. This caused it to be unable to respond quickly to small perturbations to the VCO frequency which happened when transmitting at full power. Better shielding and voltage regulation may isolate the synthesizer circuits to the point that the VCO frequency is not adversely affected, making the time constant less of an issue.

At any rate, the present loop filter value took care of most but not all of the problem. The unpredictability turned out to be related to where the rig was positioned on the desk. I didn't realize at first that strong fields from an antenna tuner/matching network were affecting the VCO frequency.

The frequency shift was most easily observed in the CW mode because it caused a chirpy CW note. The effect on SSB was more serious because it generated distortion, which made it hard for other stations to understand the speech. Separating the antenna tuner and transceiver eliminated this effect.

Power issues

One other problem caused a bit of instability. Even though I used AWG number-10 wire for the power connections, the 6-ft-long cable caused enough voltage drop to affect the VCO when going from zero to full output. This was solved by changing to two 4-ft lengths of AWG number-3 welding cable.

This may seem like overkill, but it works well. Other more sophisticated voltage regulator circuits in the synthesizer should make the synthesizer less prone to voltage changes. The larger power conductors may improve IMD levels in the final amplifier.

The wire in the welding cable is made of many fine strands to give it great flexibility. Don't plan on simply stripping the insulation and using the bare end, because it turns into a messy tangle very easily. I soldered short pieces of copper tubing over two ends, then flattened the tubing and drilled holes for screw terminals. An in-line fuse holder could be mounted here.

The other two ends connecting to the transceiver will have to connect to the short power leads coming from the transceiver. You can tin the cable ends with solder to make them a solid mass. After they cool enough, drill a hole for a small bolt and you can attach a short pigtail, fuse holder, or some other connector.

The power source should be a well-regulated supply or a "stiff" battery. If you use a battery, be sure to learn about the necessary safety precautions for the particular type. Lead-acid batteries produce explosive gases, which must not be allowed to build up.

Never set a lead-acid battery on a table beside the rig or any other place near your face when operating. I once did this with a small motorcycle-sized lead-acid battery. I thought surely such a small battery under a light load would not generate enough gases to be dangerous.

Fortunately, fragments from the hydrogen-oxygen explosion missed me and I had only to clean and neutralize the acid electrolyte before it ruined furnishings and equipment. The explosion was apparently triggered by a tiny static-electric spark when I touched the plastic battery case.

If possible, the battery should be located outside the radio room with cables led in through a fireproof conduit. Place fuses at the battery end of the cables.

Even for temporary portable applications you should at least place the battery on the floor and under a table. Wear goggles or, better yet, a full-face shield when connecting and disconnecting power leads at the battery terminals.

Overcurrent protection

The prototype transceiver uses ordinary fast-blow fuses for overcurrent protection. A separate 15-A fuse protects the power amplifier and driver, while a 2-A fuse protects the rest of the radio.

The best way to do this is as shown in figure 19-2. In the prototype, the 2-A fuse actually connects to the amplifier side of the 15-A

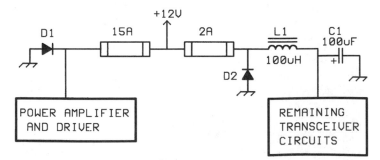

■ **19-2** *Fast-blow fuses protect the transceiver against overcurrent occurrences and diodes protect against incorrect supply polarity by causing the fuses to blow.*

fuse. This allowed me to place the large fuse externally with an in-line fuse holder; however, the remaining transceiver circuits are subjected to an added voltage drop across the fuse when the amplifier and driver draw large amounts of current.

The 100-µH inductor and 47-µF tantalum capacitor help isolate the power amplifier and driver supply line from the other circuits. D1 is a large diode to protect against reverse polarity of a power source. It should be rated at 20 A or more. A 3-A diode should suffice for D2.

If you can find Zener diodes with a 15- or 16-V rating, they will also protect the rig against overvoltage. Reverse polarity forward-biases D1 and D2, letting them take most of the current until the fuse blows.

Main program listing

Note: Remember to remove line numbers if your assembler does not allow their use.

```
 1 ;*********************************************************
 2 ;THIS IS A LIST OF THE REGISTER AND BIT NAMES
 3 ;WITH THEIR CORRESPONDING ADDRESSES. THESE MUST BE
 4 ;INCLUDED IN THE SOURCE FILE FOR THE BIT ADDRESSING
 5 ;MODES TO ASSEMBLE PROPERLY.
 6 ;
 7 ;MCS-51 INTERNAL REGISTERS
 8 ;
 9 B:       EQU      0F0H        ;B REGISTER
10 ACC:     EQU      0E0H        ;ACCUMULATOR
11 PSW:     EQU      0D0H        ;PROGRAM STATUS WORD
12 IPC:     EQU      0B8H        ;INTERRUPT PRIORITY
13 P3:      EQU      0B0H        ;PORT 3
14 IEC:     EQU      0A8H        ;INTERRUPT ENABLE
15 P2:      EQU      0A0H        ;PORT 2
16 SBUF:    EQU      99H         ;SEND BUFFER
17 SCON:    EQU      98H         ;SERIAL CONTROL
18 P1:      EQU      90H         ;PORT 1
19 TH1:     EQU      8DH         ;TIMER 1 HIGH
20 TH0:     EQU      8CH         ;TIMER 0 HIGH
21 TL1:     EQU      8BH         ;TIMER 1 LOW
22 TL0:     EQU      8AH         ;TIMER 0 LOW
23 TMOD:    EQU      89H         ;TIMER MODE
24 TCON:    EQU      88H         ;TIMER CONTROL
25 PCON:    EQU      87H         ;POWER CONTROL REGISTER
26 DPH:     EQU      83H         ;DATA POINTER HIGH
27 DPL:     EQU      82H         ;DATA POINTER LOW
28 SP:      EQU      81H         ;STACK POINTER
29 P0:      EQU      80H         ;PORT 0
30 ;
31 ;MCS-51 INTERNAL BIT ADDRESSES
32 ;
33 CY:      EQU      0D7H        ;CARRY FLAG
34 AC:      EQU      0D6H        ;AUXILIARY-CARRY FLAG
35 F0:      EQU      0D5H        ;USER FLAG 0
36 RS1:     EQU      0D4H        ;REGISTER SELECT MSB
37 RS0:     EQU      0D3H        ;REGISTER SELECT LSB
38 OV:      EQU      0D2H        ;OVERFLOW FLAG
39 P:       EQU      0D0H        ;PARITY FLAG
40 PS:      EQU      0BCH        ;PRIORITY SERIAL PORT
41 PT1:     EQU      0BBH        ;PRIORITY TIMER 1
42 PX1:     EQU      0BAH        ;PRIORITY EXTERNAL 1
43 PT0:     EQU      0B9H        ;PRIORITY TIMER 0
44 PX0:     EQU      0B8H        ;PRIORITY EXTERNAL 0
45 EA:      EQU      0AFH        ;ENABLE ALL INTERRUPT
46 ES:      EQU      0ACH        ;ENABLE SERIAL INTERRUPT
47 ET1:     EQU      0ABH        ;ENABLE TIMER 1 INTERRUPT
48 EX1:     EQU      0AAH        ;ENABLE EXTERNAL 1 INTERR
49 ET0:     EQU      0A9H        ;ENABLE TIMER 0 INTERRUPT
50 EX0:     EQU      0A8H        ;ENABLE EXTERNAL 0 INTERR
51 SM0:     EQU      09FH        ;SERIAL MODE 0
```

```
52 SM1:      EQU      09EH        ;SERIAL MODE 1
53 SM2:      EQU      09DH        ;SERIAL MODE 2
54 REN:      EQU      09CH        ;SERIAL RECEPTION ENABLE
55 TB8:      EQU      09BH        ;TRANSMIT BIT 8
56 RB8:      EQU      09AH        ;RECEIVE BIT 8
57 TI:       EQU      099H        ;TRANSMIT INTERRUPT FLAG
58 RI:       EQU      098H        ;RECEIVE INTERRUPT FLAG
59 TF1:      EQU      08FH        ;TIMER 1 OVERFLOW FLAG
60 TR1:      EQU      08EH        ;TIMER 1 RUN CONTROL BIT
61 TF0:      EQU      08DH        ;TIMER 0 OVERFLOW FLAG
62 TR0:      EQU      08CH        ;TIMER 0 RUN CONTROL BIT
63 IE1:      EQU      08BH        ;EXT INTERR. 1 EDGE FLAG
64 IT1:      EQU      08AH        ;EXT INTERR. 1 TYPE FLAG
65 IE0:      EQU      089H        ;EXT INTERR. 0 EDGE FLAG
66 IT0:      EQU      088H        ;EXT INTERR. 0 TYPE FLAG
67
68 ; ALL BIT ADDRESSES MAY BE SPECIFIED IN ONE OF THREE
69 ; WAYS:
70 ;         CLR      20H.1       ;BIT ADDRESSABLE AREA
71 ;         CLR      CY          ;BIT DIRECT ADDRESS
72 ;         CLR      PSW.7       ;BIT ADDRESSABLE REG
73
74
75 CPU "8051.TBL"
76 HEX "ON"
77 ORG 00H ;
78     AJMP 100H
79     DFS 03H-$ ; EXTERNAL INTERRUPT 0 START
80     LJMP COUNT
81     DFS 13H-$ ; EXTERNAL INTERRUPT 1 START
82     LJMP CHFAS
83     DFS 100H-$
84
85     ;START
86     CLR P3.4 ; TEMPORARY OPTO-COUPLER DISABLE
87     CLR P3.6 ; BLOCK PLL AND SHIFT REGISTER LATCHES
88     MOV R0,#0
89     MOV R1,#0                        ; **** R1 ********
90     MOV P1,#0
91 DELAY:  CLR A
92 WAIT:   INC A
93     CJNE A,#0FFH,WAIT ;
94     INC R1
95     CJNE R1,#70,DELAY ; ABOUT 36 MILLISECONDS
96     SETB P1.3 ; ENABLE HIGH
97     MOV P1,#00111000B ; FUNCTION SET DATA
98     CLR P1.3 ; ENABLE LOW
99     INC R0                           ; ****** R0 *****
100    CJNE R0,#3,DELAY ; DO FUNCTION SET 3 TIMES
101    SETB P1.3
102    CLR P1.3
103
104    ; INITIALIZATION WRITE ROUTINE
105
106    MOV R3,#0                         ; ****** R3 *****
107 FOUR:   LCALL READ
108    MOV P1,#0
109    CLR A
110    SETB P1.3 ; ENABLE HIGH
111    ACALL SUB
112    MOV R1,A
113    ANL A,#11110000B ; CLEAR LOW NIBBLE
114    ORL A,#00001000B ; KEEP ENABLE BIT SET
115    MOV P1,A ; HIGH NIBBLE DATA TO PORT 1
```

310

```
116     CLR P1.3 ; ENABLE LOW
117     MOV A,R1 ; RESTORE ACCUMULATOR
118     SWAP A ; GET LOW NIBBLE
119     ANL A,#11110000B ; LOW NIBBLE DATA, CLEAR LOW NIBBLE BITS
120     SETB P1.3
121     ORL A,#00001000B ; KEEP ENABLE BIT SET
122     MOV P1,A ; LOW NIBBLE TO PORT 1
123     CLR P1.3
124     INC R3
125     MOV A,R3
126     AJMP TEST
127 SUB:    INC A
128     ADD A,R3
129     MOVC A,@A+PC
130     RET
131     DFB 00101000B ; SYSTEM SET, 4 BIT, 2 LINES, 5 BY 7
132     DFB 00001000B ; DISPLAY OFF, NO CURSOR, NO BLINK
133     DFB 00000001B ; DISPLAY CLEAR
134     DFB 00000100B ; ENTRY MODE SET, DEC DD RAM, NO SHIFT
135     DFB 00001100B ; DISPLAY ON, CURSOR OFF, BLINK OFF
136 TEST:   CJNE A,#5,FOUR
137
138 CURC:   EQU 71H ; CURSOR (POSITION) CONTENTS
139 CURA:   EQU 70H ; CURSOR POSITION ADDRESS
140
141 NOMO:
142     CLR P3.6 ; BLOCK PLL LATCH AND SHIFT REGISTER LATCH
143
144     ; R COUNTER
145     CLR P3.0
146     SETB P3.1
147     CLR P3.1
148     CLR C
149     MOV A,#02H ; UPPER BYTE
150     RLC A ;          1
151     MOV P3.0,C
152     SETB P3.1
153     CLR P3.1
154     RLC A ;          2
155     MOV P3.0,C
156     SETB P3.1
157     CLR P3.1
158     RLC A ;          3
159     MOV P3.0,C
160     SETB P3.1
161     CLR P3.1
162     RLC A ;          4
163     MOV P3.0,C
164     SETB P3.1
165     CLR P3.1
166     RLC A ;          5
167     MOV P3.0,C
168     SETB P3.1
169     CLR P3.1
170     RLC A ;          6
171     MOV P3.0,C
172     SETB P3.1
173     CLR P3.1
174     RLC A ;          7
175     MOV P3.0,C
176     SETB P3.1
177     CLR P3.1
178     RLC A ;          8
179     MOV P3.0,C
```

```
180     SETB P3.1
181     CLR P3.1
182
183     CLR C
184     MOV A,#58H ; LOWER BYTE
185     RLC A ;          1
186     MOV P3.0,C
187     SETB P3.1
188     CLR P3.1
189     RLC A ;          2
190     MOV P3.0,C
191     SETB P3.1
192     CLR P3.1
193     RLC A ;          3
194     MOV P3.0,C
195     SETB P3.1
196     CLR P3.1
197     RLC A ;          4
198     MOV P3.0,C
199     SETB P3.1
200     CLR P3.1
201     RLC A ;          5
202     MOV P3.0,C
203     SETB P3.1
204     CLR P3.1
205     RLC A ;          6
206     MOV P3.0,C
207     SETB P3.1
208     CLR P3.1
209     RLC A ;          7
210     MOV P3.0,C
211     SETB P3.1
212     CLR P3.1
213     RLC A ;          8
214     MOV P3.0,C
215     SETB P3.1
216     CLR P3.1
217
218     SETB P3.0   ; R COUNTER LATCH SELECTED
219     SETB P3.1
220     CLR P3.1
221
222     ; SHIFT REGISTER
223     CLR C
224     MOV A,#10101010B
225     RLC A ;          1
226     MOV P3.0,C
227     SETB P3.1
228     CLR P3.1
229     RLC A ;          2
230     MOV P3.0,C
231     SETB P3.1
232     CLR P3.1
233     RLC A ;          3
234     MOV P3.0,C
235     SETB P3.1
236     CLR P3.1
237     RLC A ;          4
238     MOV P3.0,C
239     SETB P3.1
240     CLR P3.1
241     RLC A ;          5
242     MOV P3.0,C
243     SETB P3.1
```

312

```
244     CLR P3.1
245     RLC A ;         6
246     MOV P3.0,C
247     SETB P3.1
248     CLR P3.1
249     RLC A ;         7
250     MOV P3.0,C
251     SETB P3.1
252     CLR P3.1
253     RLC A ;         8
254     MOV P3.0,C
255     SETB P3.1
256     CLR P3.1
257
258     SETB P3.6 ; DATA TO PLL LATCH AND SHIFT REGISTER LATCH
259     CLR P3.6 ; BLOCK PLL LATCH AND SHIFT REGISTER LATCH
260
261 BIN:    EQU 79H
262 BIN2:   EQU 7AH
263 BIN3:   EQU 7BH
264
265     MOV IEC,#10000001B ; ENABLE ALL INTERRUPTS, TURN ON IE0,
266     SETB IT0 ; FALLING EDGE TRIGGERED
267     SETB IT1 ; FALLING EDGE TRIGGERED
268     MOV TMOD,#00001001B ; MODE 1 FOR TIMER 0, TIMER MODE,
269     ;GATED
270     SETB TR0 ; TIMER 0 SWITCHED ON
271     SETB PX1 ; INT1 HIGHEST PRIORITY
272
273
274 TMP_0:  EQU 1BH
275 TMP_1:  EQU 1CH
276 TMP_2:  EQU 1DH
277 TMP_3:  EQU 1EH
278 TMP_4:  EQU 1FH
279 OP_0:   EQU 20H ; LOW BYTE
280 OP_1:   EQU 21H
281 OP_2:   EQU 22H
282 OP_3:   EQU 23H
283 OP_4:   EQU 30H ; HI BYTE
284
285 OPTO:   EQU 78H
286 INPUTHI:  EQU 75H
287 INPUTLO:  EQU 74H
288 PLFH:   EQU 73H ; PLL VALUE, FIXED IN SPLIT MODES
289 PLFL:   EQU 72H ; PLL VALUE, FIXED IN SPLIT MODES
290       ; 24H BIT ADDRESSABLE, BIT 20H KEYLINE STATUS, BIT 21H
291     ;OUT-OF-BAND MEMORY, BIT 22H OPTO ACTIVE
292 SHRG:   EQU 25H ; SHIFT REGISTER, LOW NIB. BAND, HI NIB. MODE
293 ADLO:   EQU 26H
294 ADMID:  EQU 27H
295 ADHIER: EQU 28H
296 ADHIEST: EQU 29H
297 SUBLO: EQU ADLO
298 SUBMID: EQU ADMID
299 SUBHIER: EQU ADHIER
300 SUBHIEST: EQU ADHIEST
301
302 CTRHI:  EQU 19H ; MIXER OUTPUT FREQUENCY COUNT
303 CTRLO:  EQU 18H ; MIXER OUTPUT FREQUENCY COUNT
304 TUNLO:  EQU 2CH ; DIGITAL TO ANALOG MEMORY, TUNABLE
305 TUNHI:  EQU 2DH ; DIGITAL TO ANALOG MEMORY, TUNABLE
306 DALO:   EQU 2EH ; DIGITAL TO ANALOG MEMORY, FIXED
307 DAHI:   EQU 2FH ; DIGITAL TO ANALOG MEMORY, FIXED
```

```
308
309 PLML:   EQU 31H ; PHASE LOCK LOOP MEMORY LOW
310 PLMH:   EQU 32H ; PHASE LOCK LOOP MEMORY HIGH
311 OPTHI:  EQU 33H ; OPTO COUNTER HI BYTE
312 ; 34H IS USED AS A LOOP COUNTER FOR DIVIDE ROUTINE
313 PBM:    EQU 35H ; PUSH-BUTTON MEMORY
314 FMML:   EQU 36H  ; POINTS TO FREQUENCY-MEMORY RAM ADDRESS
315 FUNC:   EQU 37H  ; TUNE RATE HI-NIBBLE, SPLIT FUNCTIONS LOW NIBBLE
316 CALM:   EQU 38H  ; REMEMBERS LEAST BYTE OF MIXER OUT FREQ.
317 READY:  EQU 39H  ; DECREMENTED BY MAIN LOOP, ALLOWS D TO A VALUE
318         ; TO BE SET WHEN READY = 0
319     MOV OP_0,#0
320     MOV OP_1,#0
321     MOV OP_2,#0
322     MOV OP_3,#0
323     MOV OP_4,#0
324
325     MOV ADLO,#0
326     MOV ADMID,#0
327     MOV ADHIER,#0
328     MOV ADHIEST,#0
329
330 ;       MOV FUNC,#11001110B ; 10-HZ-PER-STEP TUNE RATE
331 ;       ; NORMAL/SIMPLEX MODE
332 ;       MOV FMML,#40H ; TEMPORARY INITIAL FREQ. MEMORY
333
334 ;       MOV 40H,#0A8H ; FOR TEMPORARY TESTING
335 ;       MOV 41H,#60H
336 ;       MOV 42H,#15H
337     MOV R1,#2DH ; #2DH EQUATES TO TUNHI, TEMPORARY
338     AJMP START
339 POLL:
340 ; ************************************************
341     JB P3.7,BAND ; POWER LOSS CLEARS P3.7
342     MOV P1,#0
343     MOV P3,#0
344     MOV P2,#0
345     ORL PCON,#2 ; POWER DOWN MODE, P0 HIGH IMPEDANCE
346
347     ; BAND SELECTION AND TRANSMIT FREQUENCY RANGE
348 BAND:
349     MOV R0,FMML ; MOVE FREQ. MEMORY VALUES TO TMP REGISTERS
350     MOV TMP_1,@R0
351     INC R0
352     MOV TMP_2,@R0
353     INC R0
354     MOV A,@R0
355 ANL A,#00111111B
356 MOV TMP_3,A
357
358     CLR C
359     MOV A,#40H     ; LOW BYTE OF 2.0 MHZ /10
360     SUBB A,TMP_1
361     MOV A,#0DH     ; MID BYTE OF 2.0 MHZ /10
362     SUBB A,TMP_2
363     MOV A,#3H      ; HIGH BYTE OF 2.0 MHZ /10
364     SUBB A,TMP_3
365     JC CK3.5
366     MOV TMP_4,#1
367
368     CLR C
369     MOV A,TMP_1
370     SUBB A,#20H     ;LOW BYTE OF 1.8 MHZ /10
371     MOV A,TMP_2
```

```
372      SUBB A,#0BFH    ;MID BYTE OF 1.8 MHZ /10
373      MOV A,TMP_3
374      SUBB A,#2H      ;HIGH BYTE OF 1.8 MHZ /10
375      JNC HOP1
376      SETB 21H    ; INDICATES "BELOW BAND EDGE"
377 HOP1:   AJMP DONE
378
379 CK3.5:
380      CLR C
381      MOV A,#80H      ; LOW BYTE OF 4.0 MHZ /10
382      SUBB A,TMP_1
383      MOV A,#1AH      ; MID BYTE OF 4.0 MHZ /10
384      SUBB A,TMP_2
385      MOV A,#6H       ; HIGH BYTE OF 4.0 MHZ /10
386      SUBB A,TMP_3
387      JC CK7
388      MOV TMP_4,#2
389
390      CLR C
391      MOV A,TMP_1
392      SUBB A,#30H     ;LOW BYTE OF 3.5 MHZ /10
393      MOV A,TMP_2
394      SUBB A,#57H     ;MID BYTE OF 3.5 MHZ /10
395      MOV A,TMP_3
396      SUBB A,#5H      ;HIGH BYTE OF 3.5 MHZ /10
397      JNC HOP2
398      SETB 21H    ; INDICATES "BELOW BAND EDGE"
399 HOP2:   AJMP DONE
400 CK7:
401      CLR C
402      MOV A,#90H      ; LOW BYTE OF 7.3 MHZ /10
403      SUBB A,TMP_1
404      MOV A,#23H      ; MID BYTE OF 7.3 MHZ /10
405      SUBB A,TMP_2
406      MOV A,#0BH      ; HIGH BYTE OF 7.3 MHZ /10
407      SUBB A,TMP_3
408      JC CK10
409      MOV TMP_4,#3
410
411      CLR C
412      MOV A,TMP_1
413      SUBB A,#60H     ;LOW BYTE OF 7.0 MHZ /10
414      MOV A,TMP_2
415      SUBB A,#0AEH    ;MID BYTE OF 7.0 MHZ /10
416      MOV A,TMP_3
417      SUBB A,#0AH     ;HIGH BYTE OF 7.0 MHZ /10
418      JNC HOP3
419      SETB 21H    ; INDICATES "BELOW BAND EDGE"
420 HOP3:   AJMP DONE
421 CK10:
422      CLR C
423      MOV A,#0D8H     ; LOW BYTE OF 10.15 MHZ /10
424      SUBB A,TMP_1
425      MOV A,#7CH      ; MID BYTE OF 10.15 MHZ /10
426      SUBB A,TMP_2
427      MOV A,#0FH      ; HIGH BYTE OF 10.15 MHZ /10
428      SUBB A,TMP_3
429      JC CK14
430      MOV TMP_4,#4
431
432      CLR C
433      MOV A,TMP_1
434      SUBB A,#50H     ;LOW BYTE OF 10.1 MHZ /10
435      MOV A,TMP_2
```

```
436       SUBB A,#69H     ;MID BYTE OF 10.1 MHZ /10
437       MOV A,TMP_3
438       SUBB A,#0FH     ;HIGH BYTE OF 10.1 MHZ /10
439       JNC HOP4
440       SETB 21H    ; INDICATES "BELOW BAND EDGE"
441 HOP4:    AJMP DONE
442 CK14:
443       CLR C
444       MOV A,#78H      ; LOW BYTE OF 14.35 MHZ /10
445       SUBB A,TMP_1
446       MOV A,#0E5H      ; MID BYTE OF 14.35 MHZ /10
447       SUBB A,TMP_2
448       MOV A,#15H      ; HIGH BYTE OF 14.35 MHZ /10
449       SUBB A,TMP_3
450       JC CK18
451       MOV TMP_4,#5
452
453       CLR C
454       MOV A,TMP_1
455       SUBB A,#0C0H     ;LOW BYTE OF 14.0 MHZ /10
456       MOV A,TMP_2
457       SUBB A,#5CH     ;MID BYTE OF 14.0 MHZ /10
458       MOV A,TMP_3
459       SUBB A,#15H     ;HIGH BYTE OF 14.0 MHZ /10
460       JNC HOP5
461       SETB 21H    ; INDICATES "BELOW BAND EDGE"
462 HOP5:    AJMP DONE
463 CK18:
464       CLR C
465       MOV A,#0E0H      ; LOW BYTE OF 18.168 MHZ /10
466       SUBB A,TMP_1
467       MOV A,#0B8H      ; MID BYTE OF 18.168 MHZ /10
468       SUBB A,TMP_2
469       MOV A,#1BH      ; HIGH BYTE OF 18.168 MHZ /10
470       SUBB A,TMP_3
471       JC CK21
472       MOV TMP_4,#6
473
474       CLR C
475       MOV A,TMP_1
476       SUBB A,#0D0H     ;LOW BYTE OF 18.068 MHZ /10
477       MOV A,TMP_2
478       SUBB A,#91H     ;MID BYTE OF 18.068 MHZ /10
479       MOV A,TMP_3
480       SUBB A,#1BH     ;HIGH BYTE OF 18.068 MHZ /10
481       JNC HOP6
482       SETB 21H    ; INDICATES "BELOW BAND EDGE"
483 HOP6:    AJMP DONE
484
485 CK21:
486       CLR C
487       MOV A,#0E8H      ; LOW BYTE OF 21.45 MHZ /10
488       SUBB A,TMP_1
489       MOV A,#0BAH      ; MID BYTE OF 21.45 MHZ /10
490       SUBB A,TMP_2
491       MOV A,#20H      ; HIGH BYTE OF 21.45 MHZ /10
492       SUBB A,TMP_3
493       JC CK24
494       MOV TMP_4,#7
495
496       CLR C
497       MOV A,TMP_1
498       SUBB A,#20H     ;LOW BYTE OF 21.0 MHZ /10
499       MOV A,TMP_2
```

```
500        SUBB A,#0BH    ;MID BYTE OF 21.0 MHZ /10
501        MOV A,TMP_3
502        SUBB A,#20H    ;HIGH BYTE OF 21.0 MHZ /10
503        JNC HOP7
504        SETB 21H   ; INDICATES "BELOW BAND EDGE"
505 HOP7:  AJMP DONE
506 CK24:
507        CLR C
508        MOV A,#0B8H    ; LOW BYTE OF 24.990 MHZ /10
509        SUBB A,TMP_1
510        MOV A,#21H     ; MID BYTE OF 24.990 MHZ /10
511        SUBB A,TMP_2
512        MOV A,#26H     ; HIGH BYTE OF 24.990 MHZ /10
513        SUBB A,TMP_3
514        JC CK28
515        MOV TMP_4,#8
516
517        CLR C
518        MOV A,TMP_1
519        SUBB A,#0A8H   ;LOW BYTE OF 24.890 MHZ /10
520        MOV A,TMP_2
521        SUBB A,#0FAH   ;MID BYTE OF 24.890 MHZ /10
522        MOV A,TMP_3
523        SUBB A,#25H    ;HIGH BYTE OF 24.890 MHZ /10
524        JNC HOP8
525        SETB 21H   ; INDICATES "BELOW BAND EDGE"
526 HOP8:  AJMP DONE
527 CK28:
528        CLR C
529        MOV A,#90H     ; LOW BYTE OF 29.7 MHZ /10
530        SUBB A,TMP_1
531        MOV A,#51H     ; MID BYTE OF 29.7 MHZ /10
532        SUBB A,TMP_2
533        MOV A,#2DH     ; HIGH BYTE OF 29.7 MHZ /10
534        SUBB A,TMP_3
535        JNC CK28L
536        SETB 21H ; ALSO INDICATES ABOVE 29.7 MHZ
537 CK28L:
538        MOV TMP_4,#9
539        CLR C
540        MOV A,TMP_1
541        SUBB A,#80H    ;LOW BYTE OF 28.0 MH7 /10
542        MOV A,TMP_2
543        SUBB A,#0B9H   ;MID BYTE OF 28.0 MHZ /10
544        MOV A,TMP_3
545        SUBB A,#2AH    ;HIGH BYTE OF 28.0 MHZ /10
546        JNC DONE
547        SETB 21H    ; INDICATES "BELOW BAND EDGE"
548 DONE:
549        JB P3.3,MVSR  ; IF NO XMT THEN GO AHEAD
550        JNB 21H,MVSR  ; IF ABOVE BOTTOM BAND EDGE, GO AHEAD
551        MOV TMP_4,#0  ; DISABLE XMT
552 MVSR:   MOV A,SHRG
553        ANL A,#11110000B ; CLEAR LOW NIBBLE
554        ORL A,TMP_4 ; SAVE LOW PASS FILTER SELECTION
555        MOV SHRG,A
556        CLR 21H
557        ; SAVE MODE IN FMML SELECTION ROUTINE
558
559        ; D TO A AND PLL MEMORY SELECTION
560        MOV A,FUNC
561        ANL A,#00001111B
562        CJNE A,#00000101B,TXTUN ; #00000101B = RIT
563        MOV C,P3.3
```

```
564        JC TXMD
565        MOV R1,#2DH ; #2DH EQUATES TO TUNHI
566        AJMP DODA
567 TXMD:   MOV R1,#2FH ; #2FH EQUATES TO DAHI
568        AJMP DODA
569 TXTUN:  CJNE A,#00001010B,NORMD ; #00001010B = XIT
570        MOV C,P3.3
571        JNC RXMD
572        MOV R1,#2DH ; #2DH EQUATES TO TUNHI
573        AJMP DODA
574 RXMD:   MOV R1,#2FH ; #2FH EQUATES TO DAHI
575        AJMP DODA
576 NORMD:  ; AINT XIT, AINT RIT, MUST BE NORMAL
577        MOV DALO,TUNLO
578        MOV DAHI,TUNHI
579        MOV R1,#2DH ; #2DH EQUATES TO TUNHI
580        MOV PLFL,PLML
581        MOV PLFH,PLMH
582 DODA:
583        MOV A,PBM ;
584        ANL A,#11110000B ;
585        CJNE A,#00000000B,DOTRD ; DO TUNE REGISTER DISPLAY
586        AJMP NOTRD
587 DOTRD:  MOV BIN,R5
588        MOV BIN2,R6
589        MOV A,OPTHI
590        ANL A,#00111111B
591        MOV BIN3,A
592        MOV CURA,#4FH
593        LCALL BCDR ; DISPLAY OPTO COUNTER
594 NOTRD:
595        ;      TO BE USED WITH MANUAL RANGE CONTROL CAPACITOR
596        ;      MOV C,P1.0 ; SEE IF LOOP IS PHASE LOCKED
597        ;      JC NOFMC
598        ;      LJMP DWN
599 NOFMC:
600        MOV A,TMP_0
601        CJNE A,#10110000B,BKDWN
602        ; DONT SELECT FREQ. MEMORY
603
604        MOV A,FMML
605        CLR C
606        ADD A,#3
607        MOV FMML,A
608        MOV A,#6DH
609        CLR C
610        SUBB A,FMML
611        JNC START
612        MOV FMML,#40H
613        AJMP START
614
615 BKDWN:  MOV A,TMP_0
616        CJNE A,#01110000B,DOTUN
617        ; DONT SELECT FREQ. MEMORY
618
619 DWN:    MOV A,FMML
620        CLR C
621        SUBB A,#3
622        MOV FMML,A
623        MOV A,#3EH
624        CLR C
625        SUBB A,FMML
626        JC START
627        MOV FMML,#6DH
```

Main program listing

```
628 START:
629     ; DISPLAY LETTER NAME OF FREQ. MEMORY LOCATION
630     MOV A,FMML
631     CLR C
632     SUBB A,#40H
633     MOV B,#3
634     DIV AB
635     ADD A,#01000001B ; "A" ASCII EQUIVALENT NUMBER
636     MOV CURC,A
637     MOV CURA,#0
638     LCALL SETAD
639     LCALL WRITE
640 NOLTR:
641     MOV R0,FMML ; MOVE FREQ. MEMORY VALUES TO TUNE REGISTERS
642     MOV A,@R0
643     MOV R5,A
644     INC R0
645     MOV A,@R0
646     MOV R6,A
647     INC R0
648     MOV A,@R0
649     ANL A,#00111111B
650     MOV OPTHI,A
651     AJMP CALC
652
653 DOTUN:  ; CHECK AND SAVE SPLIT FUNCTIONS,TUNE FMML IF NONE ACTIVE
654     MOV A,TMP_0
655     ANL A,#11110000B ; MASK OUT LOWER NIBBLE
656     CJNE A,#11100000B,CKRIT ; 11100000B - NORMAL
657     AJMP KPIT
658 CKRIT:  CJNE A,#10100000B,CKXIT
659     AJMP KPIT
660 CKXIT:  CJNE A,#01010000B,CKFNC
661 KPIT:   SWAP A ; ACC = 00001010B OR 00000101B NOW
662     MOV TMP_1,A ; TEMPORARY STORAGE
663     MOV A,FUNC ; UPPER NIBBLE TUNE RATE, LOWER NIBBLE SPLIT
664     ANL A,#11110000B ; MASK OUT LOWER NIBBLE
665     ORL A,TMP_1 ; LEAVE UPPER NIBBLE OF FUNC SAME AS BEFORE
666     MOV FUNC,A ; SAVE SPLIT FREQ. FUNCTION IN LOWER NIBBLE
667 CKFNC:  MOV A,FUNC
668     ANL A,#00001111B ; MASK "TUNE RATE" UPPER NIBBLE
669     CJNE A,#00001010B,NOXIT ; 00001010B = XIT
670     MOV C,P3.3
671     JNC CALC ; RECEIVE MODE = HIGH = CARRY SET
672     AJMP NOFMY ; NO FREQ. MEMORY, OPTO REGISTERS ONLY
673 NOXIT:  CJNE A,#00000101B,MVTR ; 00000101B = RIT
674     MOV C,P3.3
675     JC CALC ; XMIT MODE
676     AJMP NOFMY ; NO FREQ. MEMORY, OPTO REGISTERS ONLY
677 NOFMY:  MOV OP_0,R5
678     MOV OP_1,R6
679     MOV A,OPTHI
680     ANL A,#00111111B
681     MOV OP_2,A
682     MOV OP_3,#0
683     MOV OP_4,#0
684     AJMP NOTM
685 MVTR:   ; MOVE TUNE REGISTERS TO FMML SELECTED ADDRESSES
686
687     MOV R0,FMML ; MOVE TUNE REGISTERS TO FREQ. MEMORY ADDRESS
688     MOV A,R5
689     MOV @R0,A
690     INC R0
691     MOV A,R6
```

```
692        MOV @R0,A
693        INC R0
694        MOV A,@R0
695        ANL A,#11000000B ; CLEAR FREQUENCY BITS
696        MOV TMP_2,A ; TEMPORARY MODE BIT STORAGE
697        MOV A,OPTHI
698        ANL A,#00111111B ; LEAVE MODE BITS CLEAR
699        ORL A,TMP_2 ; MODE BITS UNCHANGED
700        MOV @R0,A ; SAVE FREQUENCY BITS
701
702        ; ********************************************
703    ; CALCULATE INTENDED MIXER OUTPUT FREQUENCY FROM DISPLAY
704    ; VALUE.
705 CALC:
706        MOV R0,FMML ; MOVE FREQ. MEMORY VALUES TO OP REGISTERS
707        MOV OP_0,@R0
708        INC R0
709        MOV OP_1,@R0
710        INC R0
711        MOV A,@R0
712        ANL A,#00111111B
713        MOV OP_2,A
714        MOV OP_3,#00H
715        MOV OP_4,#00H
716
717        MOV BIN,OP_0
718        MOV BIN2,OP_1
719        MOV BIN3,OP_2
720        MOV CURA,#0FH
721        LCALL BCDR ; DISPLAY OPTO COUNTER
722
723 NOTM:   MOV INPUTLO,#10
724        MOV INPUTHI,#0
725        LCALL MUL_16 ; 10 X DISPLAY VALUE IN OP REGISTERS
726
727        MOV A,SHRG
728        ANL A,#11110000B
729        CJNE A,#10000000B,LSB ; TEST FREQUENCY OFFSET FOR CW
730        MOV ADLO,#03CH   ; CW I.F. OFFSET LSB
731        MOV ADMID,#098H  ; CW I.F. OFFSET
732        MOV ADHIER,#04AH ; CW I.F. OFFSET
733        MOV ADHIEST,#04H ; CW I.F. OFFSET MSB
734        LJMP DOOFF
735
736 LSB:    CJNE A,#00100000B,USB ; TEST FREQUENCY OFFSET FOR LSB
737        MOV ADLO,#03CH   ; LSB I.F. OFFSET LSB
738        MOV ADMID,#098H  ; LSB I.F. OFFSET
739        MOV ADHIER,#04AH ; LSB I.F. OFFSET
740        MOV ADHIEST,#04H ; LSB I.F. OFFSET MSB
741        LJMP DOOFF
742 USB:
743        MOV ADLO,#00H    ; USB AND AM I.F. OFFSET LSB
744        MOV ADMID,#0A2H  ; USB I.F. OFFSET
745        MOV ADHIER,#04AH ; USB I.F. OFFSET
746        MOV ADHIEST,#04H ; USB I.F. OFFSET MSB
747
748 DOOFF:  LCALL ADD_32
749
750        ; EXTRACT PLL DIVISOR
751        MOV A,OP_1
752        ;MAKE OP_1 #11100000B, IGNORE OP VALUES < 2 TO 13TH
753        MOV B,#01000000B ; DECIMAL 64
754        DIV AB ; MAKE HIGHEST TWO BITS THE LOWEST BITS
755        MOV TMP_0,A ; STORE QUOTIENT
```

320

```
756        MOV A,OP_2 ;
757        MOV B,#00000100B ; DECIMAL 4
758        MUL AB ; SHIFT OP_2 TWO PLACES TO THE LEFT
759        ORL A,TMP_0 ; COMBINE PARTIAL OP_1 AND OP_2 BYTES
760        ADD A,#1
761        MOV TMP_1,#0
762        JNC NO2ND
763        INC TMP_1
764 NO2ND: MOV PLML,A ; LOW BYTE
765        MOV TMP_0,B ; SAVE HI PRODUCT
766        MOV A,OP_3
767        MOV B,#4 ; SHIFT OP_3 TWO PLACES TO THE LEFT
768        MUL AB
769        ORL A,TMP_0 ; COMBINE PARTIAL OP_2 AND OP_3 BYTES
770        ADD A,TMP_1
771        MOV PLMH,A
772 NOEXT:
773        MOV INPUTLO,#58H ; REFERENCE COUNTER DIVISOR LSB
774        MOV INPUTHI,#02H ; REFERENCE COUNTER DIVISOR MSB
775        LCALL MUL_16
776        MOV INPUTLO,PLML ; PLL DIVISOR
777        MOV INPUTHI,PLMH ; PLL DIVISOR
778        LCALL DIV_16
779
780        MOV SUBLO,OP_0
781        MOV SUBMID,OP_1
782        MOV SUBHIER,OP_2
783        MOV SUBHIEST,#0
784
785        MOV OP_0,#0E4H ; FIXED OSCILLATOR FREQ. LOW BYTE
786        MOV OP_1,#03H ; FIXED OSCILLATOR FREQ.
787        MOV OP_2,#96H ; FIXED OSCILLATOR FREQ.
788        MOV OP_3,#00H ; FIXED OSCILLATOR FREQ. HI BYTE
789        LCALL SUB_32 ; OP REGISTER HAS INTENDED MIXER OUT FREQ.
790
791        MOV A,PBM ; INTENDED MIXER OUT FREQUENCY TO BE
792        ANL A,#11110000B ; USED FOR ALIGNMENT AND
793        CJNE A,#00000000B,NOIOF ; DIAGNOSTIC PROCEDURES
794        MOV BIN,OP_0
795        MOV BIN2,OP_1
796        MOV BIN3,OP_2
797        MOV CURA,#4FH
798        LCALL BCDR ; DISPLAY INTENDED MIXER OUT FREQUENCY
799 NOIOF:
800        MOV SUBLO,OP_0 ; PREPARE FOR COMPARISON OF ACTUAL
801        MOV SUBMID,OP_1 ; MIXER OUTPUT FREQUENCY TO INTENDED
802        MOV SUBHIER,OP_2 ; MIXER OUTPUT FREQUENCY
803        MOV SUBHIEST,#0
804        ; ************************************************
805        MOV INPUTLO,2BH ; NUMBER OF PULSES
806        MOV INPUTHI,#0
807        MOV OP_0,#53H ; PROCESSOR CLOCK FREQUENCY
808        MOV OP_1,#80H ; PROCESSOR CLOCK FREQUENCY
809        MOV OP_2,#0CH ; PROCESSOR CLOCK FREQUENCY
810        MOV OP_3,#00H ; PROCESSOR CLOCK FREQUENCY
811        LCALL MUL_16
812
813        MOV INPUTLO,CTRLO
814        MOV INPUTHI,CTRHI
815        LCALL DIV_16
816        ; OP REGISTERS CONTAIN ACTUAL MIXER OUTPUT FREQUENCY
817
818        MOV TMP_0,OP_0 ; SAVE THE "OP" VALUES. THEY MUST NOT
819        MOV TMP_1,OP_1 ; BE CORRUPTED BY THE FIENDISH "PUSHUP"
```

```
820      MOV TMP_2,OP_2 ; "PULLDOWN" FREQUENCY CORRECTION CODE
821      MOV TMP_3,OP_3 ; WHEN SUBTRACTING.
822
823      MOV A,PBM ; ACTUAL MIXER OUT FREQUENCY TO BE        |
824      ANL A,#11110000B ; USED FOR ALIGNMENT AND           |
825      CJNE A,#00000000B,NOMOF ; DIAGNOSTIC PROCEDURES      |
826      MOV CURA,#47H
827      MOV BIN,OP_0
828      MOV BIN2,OP_1
829      MOV BIN3,OP_2
830      LCALL BCDR ; DISPLAY ACTUAL MIXER OUTPUT FREQUENCY
831      MOV CURA,#40H
832      LCALL SETAD
833      MOV A,#01000001B ; WRITE AN "A" TO DISPLAY
834      MOV CURC,A
835      LCALL WRITE
836
837      DEC R1
838      MOV BIN,@R1
839      INC R1
840      MOV A,@R1
841      ANL A,#00001111B
842      MOV BIN2,A
843      MOV BIN3,#0
844      MOV CURA,#07H
845      LCALL BCDR ; DISPLAY DIGI. TO ANALOG CONV. VALUE
846      ; TO BE USED FOR ALIGNMENT AND DIAGNOSTIC PROCEDURES
847      AJMP NONAM
848 NOMOF:
849    ; DISPLAY SPLIT FUNCTIONS
850      MOV A,FUNC
851      ANL A,#00001111B
852      MOV TMP_1,A ; TEMPORARY STORAGE
853      CJNE A,#00001010B,NOTXT ; NO TX TUNE / XIT
854      MOV CURA,#45H
855      LCALL SETAD
856      MOV CURC,#01011000B ; = X
857      LCALL WRITE
858      MOV CURC,#01010010B ; = R
859      LCALL WRITE
860      MOV CURA,#05H
861      LCALL SETAD
862      MOV CURC,#01011000B ; = X
863      LCALL WRITE
864      MOV CURC,#01010100B ; = T
865      LCALL WRITE
866      LJMP NONAM
867 NOTXT:   ; MAY BE RIT OR SIMPLEX
868      MOV A,TMP_1
869      CJNE A,#00000101B,NORML ;  NO RIT
870      MOV CURA,#05H
871      LCALL SETAD
872      MOV CURC,#01011000B ; = X
873      LCALL WRITE
874      MOV CURC,#01010010B ; = R
875      LCALL WRITE
876      MOV CURA,#45H
877      LCALL SETAD
878      MOV CURC,#01011000B ; = X
879      LCALL WRITE
880      MOV CURC,#01010100B ; = T
881      LCALL WRITE
882      LJMP NONAM
883 NORML:
```

```
884        MOV CURC,#00100000B ; = BLANK
885        MOV CURA,#05H
886        LCALL SETAD
887        LCALL WRITE
888        LCALL WRITE
889        MOV CURA,#45H
890        LCALL SETAD
891        LCALL WRITE
892        LCALL WRITE
893 NONAM:
894    ; ****************************************************
895    ;       MOV TUNLO,R5 ; TEMPORARY DAC INPUT
896    ;       MOV TUNHI,R6 ; TEMPORARY DAC INPUT
897    ;       LJMP NOOMO ; TEMPORARY TEST
898 ; ALLOW D TO A SERIAL DATA TRANSFER AFTER KEY-LINE LEVEL
899 ; SHIFTS, THEN SKIP COMPARISON FOR A PERIOD (READY VALUE) OF TIME.
900        MOV C,P3.3
901        JNB 20H,OOGY ; 20H IS KEY-LINE STATUS MEMORY BIT
902        CPL C
903 OOGY:   JNC RDY
904        MOV READY,#60
905 RDY:    MOV C,P3.3
906        MOV 20H,C ; SAVE STATUS OF KEY LINE
907        MOV A,READY
908        JNZ NOSER
909        AJMP SERL
910 NOSER:  DEC READY
911        LJMP NOOMO ; CHECK FREQ. AGAIN BEFORE COMPARING
912 SERL:
913
914 ; COMPARE INTENDED MIXER (OUTPUT) FREQ. TO ACTUAL MIXER FREQ.
915    ; VALUE OF INTENDED MIXER OUT FREQ. ALREADY LOADED INTO
916    ; SUB"XXXXX" REGISTERS.
917 COMP:   MOV A,OP_0
918        CJNE A,CALM,DONT
919        AJMP NOOMO
920 DONT:   MOV CALM,OP_0;
921
922 DOSA:   CLR C
923        MOV A,OP_0 ; CALCULATE SIGN
924        SUBB A,SUBLO
925        MOV A,OP_1
926        SUBB A,SUBMID
927        JC NEG ; GET SIGN (+ OR -) OF RESULT
928        MOV TMP_4,#0 ; POSITIVE RESULT INDICATOR
929        AJMP NONEG
930 NEG:    MOV TMP_4,#1 ; NEGATIVE RESULT INDICATOR
931 NONEG:
932        MOV A,TMP_4 ; SIGN INDICATOR
933        JNZ FNGD
934        CLR C
935        MOV A,OP_0
936        SUBB A,SUBLO
937        MOV TMP_1,A
938        MOV A,OP_1
939        SUBB A,SUBMID
940        MOV TMP_2,A
941        AJMP NNGD
942 FNGD:   CLR C
943        MOV A,SUBLO
944        SUBB A,OP_0
945        MOV TMP_1,A
946        MOV A,SUBMID
947        SUBB A,OP_1
```

```
 948      MOV TMP_2,A
 949 NNGD:
 950      JB 22H,MONUM ; TEST KNOB MOVEMENT BIT
 951      CLR C
 952      MOV A,TMP_2
 953      RRC A
 954      MOV TMP_2,A
 955      MOV A,TMP_1
 956      RRC A
 957      MOV TMP_1,A
 958
 959      ;   MOV A,TMP_1 ; LOW BYTE OF ADD,SUBTRACT RESULT
 960      ;   ORL A,#1
 961      ;   MOV TMP_1,A ; MAKE SURE ADD,SUBTRACT AMOUNT > ZERO
 962
 963 MONUM:   CLR 22H ; CLEAR KNOB MOVEMENT BIT
 964
 965      DEC R1 ; POINT TO LOW BYTE FIRST
 966      MOV A,TMP_4
 967      CJNE A,#1,TNUP ;
 968      CLR C
 969      MOV A,@R1
 970      SUBB A,TMP_1
 971      MOV @R1,A
 972      INC R1 ; POINT TO HIGH BYTE NEXT
 973      MOV A,@R1
 974      SUBB A,TMP_2
 975      MOV @R1,A
 976      AJMP NOOMO
 977 TNUP:    CLR C
 978      MOV A,@R1
 979      ADDC A,TMP_1
 980      MOV @R1,A
 981      INC R1 ; POINT TO HIGH BYTE NEXT
 982      MOV A,@R1
 983      ADDC A,TMP_2
 984      MOV @R1,A
 985 NOOMO:
 986      CLR EX1 ; CLEAR EXTERNAL INTERRUPT 1
 987
 988      ; A AND N COUNTER
 989
 990      CLR C
 991      MOV A,PLMH
 992      RLC A ;          1
 993      MOV P3.0,C
 994      SETB P3.1
 995      CLR P3.1
 996      RLC A ;          2
 997      MOV P3.0,C
 998      SETB P3.1
 999      CLR P3.1
1000      RLC A ;          3
1001      MOV P3.0,C
1002      SETB P3.1
1003      CLR P3.1
1004      RLC A ;          4
1005      MOV P3.0,C
1006      SETB P3.1
1007      CLR P3.1
1008      RLC A ;          5
1009      MOV P3.0,C
1010      SETB P3.1
1011      CLR P3.1
```

324

```
1012      RLC A ;          6
1013      MOV P3.0,C
1014      SETB P3.1
1015      CLR P3.1
1016      RLC A ;          7
1017      MOV P3.0,C
1018      SETB P3.1
1019      CLR P3.1
1020      RLC A ;          8
1021      MOV P3.0,C
1022      SETB P3.1
1023      CLR P3.1
1024
1025      CLR C
1026      MOV A,PLML
1027      RLC A ;          9
1028      MOV P3.0,C
1029      SETB P3.1
1030      CLR P3.1
1031      RLC A ;          10
1032      MOV P3.0,C
1033      SETB P3.1
1034      CLR P3.1
1035      CLR P3.0 ;       11    **** EMPTY SHIFT*****
1036      SETB P3.1
1037      CLR P3.1
1038      RLC A ;          12
1039      MOV P3.0,C
1040      SETB P3.1
1041      CLR P3.1
1042      RLC A ;          13
1043      MOV P3.0,C
1044      SETB P3.1
1045      CLR P3.1
1046      RLC A ;          14
1047      MOV P3.0,C
1048      SETB P3.1
1049      CLR P3.1
1050      RLC A ;          15
1051      MOV P3.0,C
1052      SETB P3.1
1053      CLR P3.1
1054      RLC A ;          16
1055      MOV P3.0,C
1056      SETB P3.1
1057      CLR P3.1
1058      RLC A ;          17
1059      MOV P3.0,C
1060      SETB P3.1
1061      CLR P3.1
1062
1063      CLR P3.0 ; N AND A COUNTER LATCH SELECTED
1064      SETB P3.1
1065      CLR P3.1
1066
1067      ; SHIFT REGISTER
1068      CLR C
1069      MOV A,SHRG
1070      RLC A ;          1
1071      MOV P3.0,C
1072      SETB P3.1
1073      CLR P3.1
1074      RLC A ;          2
1075      MOV P3.0,C
```

```
1076        SETB P3.1
1077        CLR P3.1
1078        RLC A ;          3
1079        MOV P3.0,C
1080        SETB P3.1
1081        CLR P3.1
1082        RLC A ;          4
1083        MOV P3.0,C
1084        SETB P3.1
1085        CLR P3.1
1086        RLC A ;          5
1087        MOV P3.0,C
1088        SETB P3.1
1089        CLR P3.1
1090        RLC A ;          6
1091        MOV P3.0,C
1092        SETB P3.1
1093        CLR P3.1
1094        RLC A ;          7
1095        MOV P3.0,C
1096        SETB P3.1
1097        CLR P3.1
1098        RLC A ;          8
1099        MOV P3.0,C
1100        SETB P3.1
1101        CLR P3.1
1102
1103        SETB P3.6 ; DATA TO SHIFT REGISTER LATCH
1104        CLR P3.6 ; BLOCK SHIFT REGISTER LATCH
1105
1106        ; D TO A
1107
1108        CLR C
1109        MOV A,@R1 ; HIGH BYTE FIRST
1110        SWAP A
1111        ANL A,#11110000B
1112        CPL A
1113        CLR C
1114        RLC A ;          1
1115        MOV P3.0,C
1116        SETB P3.1
1117        CLR P3.1
1118        RLC A ;          2
1119        MOV P3.0,C
1120        SETB P3.1
1121        CLR P3.1
1122        RLC A ;          3
1123        MOV P3.0,C
1124        SETB P3.1
1125        CLR P3.1
1126        RLC A ;          4
1127        MOV P3.0,C
1128        SETB P3.1
1129        CLR P3.1
1130
1131        DEC R1
1132        CLR C
1133        MOV A,@R1 ; LOW BYTE FOR D TO A CONVERTER
1134
1135        CPL A
1136        RLC A ;          5
1137        MOV P3.0,C
1138        SETB P3.1
1139        CLR P3.1
```

```
1140        RLC A ;          6
1141        MOV P3.Ø,C
1142        SETB P3.1
1143        CLR P3.1
1144        RLC A ;          7
1145        MOV P3.Ø,C
1146        SETB P3.1
1147        CLR P3.1
1148        RLC A ;          8
1149        MOV P3.Ø,C
1150        SETB P3.1
1151        CLR P3.1
1152        RLC A ;          9
1153        MOV P3.Ø,C
1154        SETB P3.1
1155        CLR P3.1
1156        RLC A ;          10
1157        MOV P3.Ø,C
1158        SETB P3.1
1159        CLR P3.1
1160        RLC A ;          11
1161        MOV P3.Ø,C
1162        SETB P3.1
1163        CLR P3.1
1164        RLC A ;          12
1165        MOV P3.Ø,C
1166        SETB P3.1
1167        CLR P3.1
1168
1169        CLR P3.7 ; DATA TO D-A LATCH
1170        SETB P3.7 ; BLOCK D-A LATCH
1171        INC R1 ; RESTORE R1 TO PREVIOUS VALUE
1172        SETB EX1 ; ENABLE EXTERNAL INTERRUPT 1
1173
1174    ;     MOV CURA,#Ø7H
1175    ;     MOV BIN,PLML
1176    ;     MOV BIN2,PLMH
1177    ;     LCALL BCDR ; DISPLAY PLL DIVISOR
1178
1179 PBR:     ; PUSH-BUTTON ROUTINE
1180        MOV TMP_Ø,#Ø
1181        MOV TMP_1,#Ø
1182        MOV TMP_2,#Ø
1183        MOV TMP_3,#Ø
1184        CLR P3.5 ; ENABLE PUSH-BUTTONS
1185 GOBCK: MOV A,P1
1186        ANL A,#1111ØØØØB ; BUTTONS ONLY
1187        CJNE A,TMP_Ø,DIFER
1188        LJMP SVBUT
1189 DIFER: MOV TMP_1,#Ø
1190        MOV TMP_2,#Ø
1191 SVBUT:
1192        MOV TMP_Ø,A ; SAVE BUTTON VALUE
1193        CJNE A,#1111ØØØØB,TIMEO
1194        LJMP KEEP
1195 TIMEO:
1196        CLR C
1197        MOV A,TMP_1
1198        ADDC A,#1
1199        MOV TMP_1,A
1200        MOV A,TMP_2
1201        ADDC A,#Ø
1202        MOV TMP_2,A
1203        CJNE A,#30,GOBCK
```

```
1204       MOV A,TMP_0
1205       LJMP GOAHD
1206 KEEP:   MOV A,PBM
1207 GOAHD:
1208       MOV PBM,A
1209 NOBUT:  SETB P3.5 ; DISABLE  PUSH-BUTTONS
1210
1211    ; TUNING RATE SELECTION
1212
1213       MOV A,TMP_0
1214       CJNE A,#11010000B,NOTRC
1215
1216       MOV A,FUNC
1217       SWAP A ; PUT TUNE RATES IN LOW NIBBLE
1218       SUBB A,#1
1219       ANL A,#00001111B ; KEEP SPLITS NIBBLE CLEAR
1220       SWAP A ; RESTORE NIBBLE ORDER
1221       MOV TMP_1,A ; TEMPORARY STORAGE OF NEW TUNE RATE NIBBLE
1222       MOV A,FUNC
1223       ANL A,#00001111B ; CLEAR TUNE RATES NIBBLE, KEEP SPLIT
1224       ORL A,TMP_1 ; INSERT NEW TUNE RATE
1225       MOV FUNC,A ; STORE HIGH NIBBLE
1226
1227 NOTRC:
1228       MOV A,FUNC
1229       ANL A,#00110000B ; LIMIT UPPER NIBBLE TO MODULO 4
1230
1231
1232       CJNE A,#00010000B,NOMED
1233       MOV TMP_1,#10
1234       MOV TMP_2,#0
1235       MOV TMP_3,#0
1236       MOV CURC,#01101101B ; "m"
1237 NOMED:  CJNE A,#00100000B,NOFAST
1238       MOV TMP_1,#0E8H
1239       MOV TMP_2,#03H
1240       MOV TMP_3,#0
1241       MOV CURC,#01100110B ; "f"
1242 NOFAST: CJNE A,#00110000B,NOMEG
1243       MOV TMP_1,#0A0H
1244       MOV TMP_2,#86H
1245       MOV TMP_3,#01H
1246       MOV CURC,#01101000B ; "h"
1247 NOMEG:  CJNE A,#00000000B,NOSLOW
1248       MOV TMP_1,#1
1249       MOV TMP_2,#0
1250       MOV TMP_3,#0
1251       MOV CURC,#01110011B ; "s"
1252 NOSLOW:
1253       MOV CURA,#08H
1254       LCALL SETAD
1255       LCALL WRITE
1256
1257
1258    ; OPTO INTERRUPTER ROUTINE
1259       SETB P3.4 ; ENABLE OPTOCOUPLER
1260       MOV A,P1
1261       ANL A,#11000000B
1262       MOV R4,A ; R4 NOW HOLDS PORT DATA   **** R4 *****
1263       MOV A,OPTO
1264       CJNE A,#01000000B,CK11B
1265       CJNE R4,#00000000B,P11B
1266       AJMP DECR
1267 P11B:   CJNE R4,#11000000B,NOCT
```

328

```
1268      AJMP INCR
1269 CK11B:  MOV A,OPTO
1270      CJNE A,#11000000B,CK10B
1271      CJNE R4,#01000000B,P10B
1272      AJMP DECR
1273 P10B:   CJNE R4,#10000000B,NOCT
1274      AJMP INCR
1275 CK10B:  MOV A,OPTO
1276      CJNE A,#10000000B,M00B
1277      CJNE R4,#11000000B,P00B
1278      AJMP DECR
1279 P00B:   CJNE R4,#00000000B,NOCT
1280      AJMP INCR
1281 M00B:   CJNE R4,#10000000B,P01B
1282      AJMP DECR
1283 P01B:   CJNE R4,#01000000B,NOCT
1284
1285 INCR:   CLR C
1286      MOV A,R5
1287      ADDC A,TMP_1
1288      MOV R5,A
1289      MOV A,R6
1290      ADDC A,TMP_2
1291      MOV R6,A
1292      MOV A,OPTHI
1293      ADDC A,TMP_3
1294      MOV OPTHI,A
1295      LJMP KKTUN
1296
1297 DECR:   CLR C
1298      MOV A,R5
1299      SUBB A,TMP_1
1300      MOV R5,A
1301      MOV A,R6
1302      SUBB A,TMP_2
1303      MOV R6,A
1304      MOV A,OPTHI
1305      SUBB A,TMP_3
1306      MOV OPTHI,A
1307 KKTUN:  SETB 22H ; SET KNOB MOVEMENT BIT
1308 NOCT:   MOV OPTO,R4
1309      CLR P3.4 ; DISABLE OPTOCOUPLER
1310
1311      ; SAVE MODE
1312      MOV A,TMP_0 ; CURRENT BUTTON VALUE
1313      CJNE A,#11000000B,NOCW ; TMP_0 = #11000000B = CW
1314      MOV TMP_2,#01000000B
1315      AJMP SAVM
1316 NOCW:   CJNE A,#00110000B,NOUSB ; TMP_0 = #00110000B = USB
1317      MOV TMP_2,#11000000B
1318      AJMP SAVM
1319 NOUSB:  CJNE A,#10010000B,MODE4 ; TMP_0 = #10010000B = LSB
1320      MOV TMP_2,#10000000B
1321      AJMP SAVM
1322 MODE4:  CJNE A,#01100000B,NOSAV ; TMP_0 = #01100000B = MODE4 ? AM
1323      MOV TMP_2,#00000000B
1324
1325 SAVM:
1326      MOV R0,FMML ; POINT TO FREQ. MEMORY LSB
1327      INC R0
1328      INC R0
1329      MOV A,@R0 ; FREQUENCY MEMORY MSB
1330      ANL A,#00111111B ; CLEAR MODE BITS, FREQ. BITS UNCHANGED
1331      ORL A,TMP_2 ; NEW MODE BITS
```

329

```
1332      MOV @R0,A ; SAVE MODE AND FREQUENCY BITS
1333 NOSAV:
1334      MOV CURA,#42H
1335      LCALL SETAD ; KEEP ACC. INTACT, SET ADDRESS HERE
1336      ; SELECT MODE
1337      MOV R0,FMML ; MOVE FREQ. MEMORY MSB TO ACC
1338      INC R0
1339      INC R0
1340      MOV A,@R0
1341      ANL A,#11000000B ; CLEAR FREQUENCY BITS
1342
1343      CJNE A,#01000000B,CWNO
1344      MOV TMP_1,#10000000B
1345      MOV CURC,#00100000B    ; BLANK
1346      LCALL WRITE
1347      MOV CURC,#01010111B    ; "W"
1348      LCALL WRITE
1349      MOV CURC,#01000011B    ; "C"
1350      LCALL WRITE
1351      AJMP SAVSR
1352 CWNO:
1353      CJNE A,#11000000B,USBNO
1354      MOV TMP_1,#01000000B
1355      MOV CURC,#01000010B    ; "B"
1356      LCALL WRITE
1357      MOV CURC,#01010011B    ; "S"
1358      LCALL WRITE
1359      MOV CURC,#01010101B    ; "U"
1360      LCALL WRITE
1361      AJMP SAVSR
1362 USBNO:
1363      CJNE A,#10000000B,LSBNO
1364      MOV TMP_1,#00100000B
1365      MOV CURC,#01000010B    ; "B"
1366      LCALL WRITE
1367      MOV CURC,#01010011B    ; "S"
1368      LCALL WRITE
1369      MOV CURC,#01001100B    ; "L"
1370      LCALL WRITE
1371      AJMP SAVSR
1372 LSBNO:
1373      ; MUST BE MODE4 AM ?
1374      MOV TMP_1,#00010000B
1375      MOV CURC,#00100000B    ; BLANK
1376      LCALL WRITE
1377      MOV CURC,#01001101B    ; "M"
1378      LCALL WRITE
1379      MOV CURC,#01000001B    ; "A"
1380      LCALL WRITE
1381 SAVSR:
1382      MOV A,SHRG
1383      ANL A,#00001111B ; CLEAR MODE BITS, BAND BITS UNCHANGED
1384      ORL A,TMP_1 ; LOW NIBBLE UNCHANGED, NEW MODE BITS
1385      MOV SHRG,A ; STORE HIGH NIBBLE
1386
1387      LJMP POLL
1388
1389      ; MAKE BCD
1390 BCD:    EQU 7CH
1391 BCD2:   EQU 7DH
1392 BCD3:   EQU 7EH
1393 BCD4:   EQU 7FH
1394
1395 BCDR:   MOV R2,#0                          ; ****** R2 *****
```

```
1396      MOV BCD,#0
1397      MOV BCD2,#0
1398      MOV BCD3,#0
1399      MOV BCD4,#0
1400 REPEAT: MOV A,BIN
1401      CLR C
1402      RLC A ; BIN * 2
1403      MOV BIN,A
1404      MOV A,BIN2
1405      RLC A
1406      MOV BIN2,A
1407      MOV A,BIN3
1408      RLC A
1409      MOV BIN3,A
1410      MOV A,BCD
1411      ADDC A,BCD ; BCD * 2 + CARRY
1412      DAA
1413      MOV BCD,A
1414      MOV A,BCD2
1415      ADDC A,BCD2
1416      DAA
1417      MOV BCD2,A
1418      MOV A,BCD3
1419      ADDC A,BCD3
1420      DAA
1421      MOV BCD3,A
1422      MOV A,BCD4
1423      ADDC A,BCD4
1424      DAA
1425      MOV BCD4,A
1426      INC R2
1427      CJNE R2,#24,REPEAT
1428
1429      ; DISPLAY DECIMAL NUMERALS
1430
1431      LCALL SETAD
1432      MOV A,BCD ; ONES
1433      ANL A,#00001111B
1434      ORL A,#00110000B
1435      MOV CURC,A
1436      LCALL WRITE
1437      MOV A,BCD
1438      SWAP A ; TENS
1439      ANL A,#00001111B
1440      ORL A,#00110000B
1441      MOV CURC,A
1442      LCALL WRITE
1443
1444      MOV A,BCD2 ; HUNDREDS
1445      ANL A,#00001111B
1446      ORL A,#00110000B
1447      MOV CURC,A
1448      LCALL WRITE
1449      MOV A,BCD2
1450      SWAP A   ; THOUSANDS
1451      ANL A,#00001111B
1452      ORL A,#00110000B
1453      MOV CURC,A
1454      LCALL WRITE
1455
1456      MOV A,BCD3 ; TEN THOUSANDS
1457      ANL A,#00001111B
1458      ORL A,#00110000B
1459      MOV CURC,A
```

331

```
1460      LCALL WRITE
1461      MOV A,BCD3
1462      SWAP A  ; HUNDRED THOUSANDS
1463      ANL A,#00001111B
1464      ORL A,#00110000B
1465      MOV CURC,A
1466      LCALL WRITE
1467
1468      MOV A,BCD4  ; MILLIONS
1469      ANL A,#00001111B
1470      ORL A,#00110000B
1471      MOV CURC,A
1472      LCALL WRITE
1473      RET
1474
1475 SETAD:  LCALL READ
1476      MOV P1,#0 ; "SET DD RAM ADDRESS COMMAND"
1477      SETB P1.3 ; ENABLE HIGH
1478      MOV A,CURA ; " " " " ACCUMULATOR
1479      ANL A,#11110000B ; CLEAR LOW NIBBLE
1480      ORL A,#10001000B ; KEEP ENABLE BIT SET, SET DD ADR.
1481      MOV P1,A ; HIGH NIBBLE DATA TO PORT 1
1482      CLR P1.3 ; ENABLE LOW
1483      MOV A,CURA ; RESTORE ACCUMULATOR
1484      SWAP A ; GET LOW NIBBLE
1485      ANL A,#11110000B ; GET LOW DATA, CLEAR LOW NIBBLE
1486      SETB P1.3
1487      ORL A,#00001000B ; KEEP ENABLE BIT SET
1488      MOV P1,A ; LOW NIBBLE TO PORT 1
1489      CLR P1.3
1490      RET
1491
1492 WRITE:  LCALL READ
1493
1494      ; WRITE TO DD RAM
1495      MOV P1,#00000011B ; WRITE TO DD RAM, P1.0 NOT USED
1496      SETB P1.3 ; ENABLE HIGH
1497      MOV A,CURC ; CHARACTER IN CPU RAM
1498      ANL A,#11110000B ; CLEAR LOW NIBBLE
1499      ORL A,#00001011B ; KEEP ENABLE BIT SET
1500      MOV P1,A ; HIGH NIBBLE DATA TO PORT 1
1501      CLR P1.3 ; ENABLE LOW
1502      MOV A,CURC ; RESTORE ACCUMULATOR
1503      SWAP A ; GET LOW NIBBLE
1504      ANL A,#11110000B ; GET LOW DATA, CLEAR LOW NIBBLE
1505      SETB P1.3
1506      ORL A, #00001011B ; KEEP ENABLE BIT SET
1507      MOV P1,A ; LOW NIBBLE TO PORT 1
1508      CLR P1.3
1509      MOV P1,#11110111B ; THESE BITS HIGH FOR OTHER ROUTINES
1510      RET
1511
1512 READ:   MOV P1,#10000100B ; ALLOW READ, R/W HIGH
1513      CLR C
1514      SETB P1.3 ; ENABLE HIGH
1515      MOV C,P1.7 ; READ BUSY FLAG
1516      CLR P1.3 ; ENABLE LOW
1517      SETB P1.3 ; ENABLE HIGH
1518      CLR P1.3
1519      JC READ ; JUMP IF BUSY
1520 RET
1521
1522 COUNT:  PUSH PSW
1523      PUSH ACC
```

332

```
1524      PUSH B
1525      PUSH DPL
1526      PUSH DPH
1527      INC 2AH
1528      CLR C
1529      MOV A,TLØ
1530      ADDC A,#255
1531      MOV A,THØ
1532      ADDC A,#10000111B ; STORE WHEN COUNTER >= 30721D
1533      JNC POOP
1534      MOV CTRLO,TLØ
1535      MOV CTRHI,THØ
1536      MOV 2BH,2AH
1537      MOV TLØ,#0
1538      MOV THØ,#0
1539      MOV 2AH,#0
1540 POOP:   POP DPH
1541      POP DPL
1542      POP B
1543      POP ACC
1544      POP PSW
1545      RETI
1546
1547 INLO:   EQU 76H
1548 INHI:   EQU 77H
1549 DIV_16:
1550        ; THIS DIVIDES THE 40 BIT OP REGISTER BY THE VALUE
1551        ; SUPPLIED
1552      MOV R7,#0 ; ZERO OUT PARTIAL REMAINDER, HI BYTE
1553      MOV R3,#0 ; ZERO OUT PARTIAL REMAINDER, LEAST BYTE
1554      MOV TMP_0,#0 ; LEAST BYTE OF DIVIDEND
1555      MOV TMP_1,#0 ; BYTE OF DIVIDEND
1556      MOV TMP_2,#0
1557      MOV TMP_3,#0
1558      MOV TMP_4,#0 ; HI BYTE OF DIVIDEND
1559      MOV INHI,INPUTHI
1560      MOV INLO,INPUTLO
1561      MOV 34H,#40 ; LOOP COUNT
1562 DIV_LOOP:
1563      ; SHIFT THE DIVIDEND ONE BIT TO THE LEFT AND RETURN THE
1564      ; MSB IN C
1565      CLR C
1566      MOV A,OP_Ø
1567      RLC A
1568      MOV OP_Ø,A
1569      MOV A,OP_1
1570      RLC A
1571      MOV OP_1,A
1572      MOV A,OP_2
1573      RLC A
1574      MOV OP_2,A
1575      MOV A,OP_3
1576      RLC A
1577      MOV OP_3,A
1578      MOV A,OP_4
1579      RLC A
1580      MOV OP_4,A
1581
1582      MOV A,R3 ; SHIFT CARRY INTO LSB OF PARTIAL REMAINDER
1583      RLC A
1584      MOV R3,A
1585      MOV A,R7
1586      RLC A
1587      MOV R7,A
```

```
1588      ; NOW TEST TO SEE IF R7:R3 >= INHI:INLO
1589      CLR C
1590      MOV A,R7  ; SUBTRACT INHI FROM R7 TO SEE IF INHI<R7
1591      SUBB A,INHI ; A=R7-INHI, CARRY SET IF R7<INHI
1592      JC CANT_SUB
1593      ;AT THIS POINT R7>INHI OR R7=INHI
1594      JNZ CAN_SUB ; JUMP IF R7>INHI
1595      ; IF R7=INHI, TEST FOR R3>=INLO
1596      CLR C
1597      MOV A,R3
1598      SUBB A,INLO ; A=R3-INLO, CARRY SET IF R3 < INLO
1599      JC CANT_SUB
1600 CAN_SUB:
1601      ;SUBTRACT THE DIVISOR FROM THE PARTIAL REMAINDER
1602      CLR C
1603      MOV A,R3
1604      SUBB A,INLO ; A=R3-INLO
1605      MOV R3,A
1606      MOV A,R7
1607      SUBB A,INHI ; A=R7-INHI-BORROW
1608      MOV R7,A
1609      SETB C ; SHIFT A 1 INTO THE QUOTIENT
1610      LJMP QUOT
1611 CANT_SUB:
1612      ;SHIFT A 0 INTO THE QUOTIENT
1613      CLR C
1614 QUOT:
1615      ; SHIFT THE CARRY BIT INTO THE QUOTIENT
1616      ; SHIFT_Q
1617      ; SHIFT THE QUOTIENT ONE BIT TO THE LEFT AND SHIFT THE C
1618      ; INTO LSB
1619      MOV A,TMP_0
1620      RLC A
1621      MOV TMP_0,A
1622      MOV A,TMP_1
1623      RLC A
1624      MOV TMP_1,A
1625      MOV A,TMP_2
1626      RLC A
1627      MOV TMP_2,A
1628      MOV A,TMP_3
1629      RLC A
1630      MOV TMP_3,A
1631      MOV A,TMP_4
1632      RLC A
1633      MOV TMP_4,A
1634
1635      ; TEST FOR COMPLETION
1636      DJNZ 34H,DIV_LOOP
1637      ; NOW WE ARE ALL DONE, MOVE THE TMP VALUES BACK INTO OP
1638      MOV OP_0,TMP_0
1639      MOV OP_1,TMP_1
1640      MOV OP_2,TMP_2
1641      MOV OP_3,TMP_3
1642      MOV OP_4,TMP_4
1643      RET
1644
1645 MUL_16:
1646      ; MULTIPLY THE 40 BIT OP WITH THE 16 BIT VALUE SUPPLIED
1647      MOV TMP_4,#0 ; CLEAR OUT UPPER 24 BITS
1648      MOV TMP_3,#0
1649      MOV TMP_2,#0
1650      ; GENERATE THE LOWEST BYTE OF THE RESULT
1651      MOV B,OP_0
```

```
1652        MOV A,INPUTLO
1653        MUL AB
1654        MOV TMP_0,A ; LOW ORDER RESULT
1655        MOV TMP_1,B ; HIGH ORDER RESULT
1656        ; NOW GENERATE THE NEXT HIGHER ORDER BYTE
1657        MOV B,OP_1
1658        MOV A,INPUTLO
1659        MUL AB
1660        ADD A,TMP_1 ; LOW ORDER RESULT
1661        MOV TMP_1,A ; SAVE
1662        MOV A,B ; GET HIGH ORDER RESULT
1663        ADDC A,TMP_2 ; INCLUDE CARRY FROM PREVIOUS OPERATION
1664        MOV TMP_2,A ; SAVE
1665        JNC MUL_LOOP0
1666        INC TMP_3
1667        ; ************************************
1668 MUL_LOOP0:
1669        MOV B,OP_2
1670        MOV A,INPUTLO
1671        MUL AB
1672        ADD A,TMP_2 ; LOW ORDER RESULT
1673        MOV TMP_2,A ; SAVE
1674        MOV A,B ;  GET HIGH ORDER RESULT
1675        ADDC A,TMP_3 ; INCLUDE CARRY FROM PREVIOUS OPERATION
1676        MOV TMP_3,A ; SAVE
1677
1678        JNC MUL_LOOP1
1679        INC TMP_4 ; PROPAGATE CARRY INTO TMP_4
1680 MUL_LOOP1:
1681        MOV B,OP_0
1682        MOV A,INPUTHI
1683        MUL AB
1684        ADD A,TMP_1 ; LOW ORDER RESULT
1685        MOV TMP_1,A ; SAVE
1686        MOV A,B ; GET HIGH ORDER RESULT
1687        ADDC A,TMP_2 ; INCLUDE CARRY FROM PREVIOUS OPERATION
1688        MOV TMP_2,A ; SAVE
1689        JNC NOADDCY
1690        INC TMP_3
1691
1692 NOADDCY:
1693        MOV B,OP_1
1694        MOV A,INPUTHI
1695        MUL AB
1696        ADD A,TMP_2 ; LOW ORDER RESULT
1697        MOV TMP_2,A ; SAVE
1698        MOV A,B ; GET HIGH ORDER RESULT
1699        ADDC A,TMP_3 ; INCLUDE CARRY FROM PREVIOUS OPERATION
1700        MOV TMP_3,A ; SAVE
1701
1702        JNC MUL_LOOP2
1703        INC TMP_4 ; PROPAGATE CARRY INTO TMP_4
1704
1705 MUL_LOOP2:
1706        ; NOW START WORKING ON THE FOURTH BYTE
1707
1708        MOV B,OP_2
1709        MOV A,INPUTHI
1710        MUL AB
1711        ADD A,TMP_3 ; LOW ORDER RESULT
1712        MOV TMP_3,A ; SAVE
1713        MOV A,B ; GET HIGH ORDER RESULT
1714        ADDC A,TMP_4 ; INCLUDE CARRY FROM PREVIOUS OPERATION
1715        MOV  TMP_4,A ; SAVE
```

335

```
1716
1717        ; NOW THE OTHER TWO-FIFTHS
1718
1719        MOV B,OP_3
1720        MOV A,INPUTLO
1721        MUL AB
1722        ADD A,TMP_3 ; LOW ORDER RESULT
1723        MOV TMP_3,A ; SAVE
1724        MOV A,B ; GET HIGH ORDER RESULT
1725        ADDC A,TMP_4 ; INCLUDE CARRY FROM PREVIOUS OPERATION
1726        MOV TMP_4,A ; SAVE
1727
1728
1729        ; NOW FINISH OFF THE HIGHEST ORDER BYTE
1730        MOV B,OP_4
1731        MOV A,INPUTLO
1732        MUL AB
1733        ADD A,TMP_4 ; LOW ORDER RESULT
1734        MOV TMP_4,A ; SAVE
1735
1736        ; FORGET ABOUT THE HIGH ORDER RESULT, THIS IS ONLY 40 BIT MATH!
1737        MOV B,OP_3
1738        MOV A,INPUTHI
1739        MUL AB
1740        ADD A,TMP_4 ; LOW ORDER RESULT
1741        MOV TMP_4,A ; SAVE
1742        ; NOW WE ARE ALL DONE, MOVE THE TMP VALUES BACK INTO OP
1743        MOV OP_0,TMP_0
1744        MOV OP_1,TMP_1
1745        MOV OP_2,TMP_2
1746        MOV OP_3,TMP_3
1747        MOV OP_4,TMP_4
1748        RET
1749
1750 ADD_32:
1751        ; ADD THE 32 BITS SUPPLIED BY THE CALLER TO THE OP
1752        ; REGISTERS
1753        CLR C
1754        MOV A,OP_0
1755        ADDC A,ADLO ; LOWEST BYTE FIRST
1756        MOV OP_0,A
1757        MOV A,OP_1
1758        ADDC A,ADMID ; MID-LOWEST BYTE + CARRY
1759        MOV OP_1,A
1760        MOV A,OP_2
1761        ADDC A,ADHIER ; MID-HIGHEST BYTE + CARRY
1762        MOV OP_2,A
1763        MOV A,OP_3
1764        ADDC A,ADHIEST ; HIGHEST BYTE + CARRY
1765        MOV OP_3,A
1766        RET
1767 SUB_32:
1768        ; SUBTRACT THE 32 BITS SUPPLIED BY THE CALLER FROM THE OP
1769        ; REGISTERS
1770        CLR C
1771        MOV A,OP_0
1772        SUBB A,SUBLO ; LOW BYTE FIRST
1773        MOV OP_0,A
1774        MOV A,OP_1
1775        SUBB A,SUBMID ; MID-LOWEST BYTE + CARRY
1776        MOV OP_1,A
1777        MOV A,OP_2
1778        SUBB A,SUBHIER ; MID-HIGHER BYTE + CARRY
1779        MOV OP_2,A
```

336

```
1780        MOV A,OP_3
1781        SUBB A,SUBHIEST ; HIGHEST BYTE + CARRY
1782        MOV OP_3,A
1783        RET
1784
1785 CHFAS:  CLR EX1 ; CLEAR EXTERNAL INTERRUPT 1
1786      PUSH PSW
1787      PUSH ACC
1788      PUSH B
1789      PUSH DPL
1790      PUSH DPH
1791      PUSH 1H ; 1H = R1
1792
1793      ; PLL MEMORY SELECTION
1794      MOV A,FUNC
1795      ANL A,#00001111B
1796      CJNE A,#00000101B,NOFX ; #00000101B = RIT, FIX XMT FREQ.
1797      MOV R1,#73H ; #73H EQUATES TO PLFH
1798      LJMP DOPL
1799 NOFX:   ; AINT RIT, MAY BE XIT OR NORMAL, #00001010B = XIT
1800      MOV R1,#32H ; #32H EQUATES TO PLMH
1801 DOPL:
1802
1803      ; A AND N COUNTER
1804      CLR C
1805      MOV A,@R1 ; HIGH BYTE FIRST
1806      RLC A ;       1
1807      MOV P3.0,C
1808      SETB P3.1
1809      CLR P3.1
1810      RLC A ;       2
1811      MOV P3.0,C
1812      SETB P3.1
1813      CLR P3.1
1814      RLC A ;       3
1815      MOV P3.0,C
1816      SETB P3.1
1817      CLR P3.1
1818      RLC A ;       4
1819      MOV P3.0,C
1820      SETB P3.1
1821      CLR P3.1
1822      RLC A ;       5
1823      MOV P3.0,C
1824      SETB P3.1
1825      CLR P3.1
1826      RLC A ;       6
1827      MOV P3.0,C
1828      SETB P3.1
1829      CLR P3.1
1830      RLC A ;       7
1831      MOV P3.0,C
1832      SETB P3.1
1833      CLR P3.1
1834      RLC A ;       8
1835      MOV P3.0,C
1836      SETB P3.1
1837      CLR P3.1
1838
1839      CLR C
1840      DEC R1
1841      MOV A,@R1 ; NEXT, LOW BYTE
1842      RLC A ;       9
1843      MOV P3.0,C
```

```
1844        SETB P3.1
1845        CLR P3.1
1846        RLC A ;           10
1847        MOV P3.0,C
1848        SETB P3.1
1849        CLR P3.1
1850        CLR P3.0 ;        11   **** EMPTY SHIFT*****
1851        SETB P3.1
1852        CLR P3.1
1853        RLC A ;           12
1854        MOV P3.0,C
1855        SETB P3.1
1856        CLR P3.1
1857        RLC A ;           13
1858        MOV P3.0,C
1859        SETB P3.1
1860        CLR P3.1
1861        RLC A ;           14
1862        MOV P3.0,C
1863        SETB P3.1
1864        CLR P3.1
1865        RLC A ;           15
1866        MOV P3.0,C
1867        SETB P3.1
1868        CLR P3.1
1869        RLC A ;           16
1870        MOV P3.0,C
1871        SETB P3.1
1872        CLR P3.1
1873        RLC A ;           17
1874        MOV P3.0,C
1875        SETB P3.1
1876        CLR P3.1
1877
1878        CLR P3.0 ; N AND A COUNTER LATCH SELECTED
1879        SETB P3.1
1880        CLR P3.1
1881
1882        ; SHIFT REGISTER
1883        CLR C
1884        MOV A,SHRG
1885        RLC A ;           1
1886        MOV P3.0,C
1887        SETB P3.1
1888        CLR P3.1
1889        RLC A ;           2
1890        MOV P3.0,C
1891        SETB P3.1
1892        CLR P3.1
1893        RLC A ;           3
1894        MOV P3.0,C
1895        SETB P3.1
1896        CLR P3.1
1897        RLC A ;           4
1898        MOV P3.0,C
1899        SETB P3.1
1900        CLR P3.1
1901        RLC A ;           5
1902        MOV P3.0,C
1903        SETB P3.1
1904        CLR P3.1
1905        RLC A ;           6
1906        MOV P3.0,C
1907        SETB P3.1
```

338

```
1908        CLR P3.1
1909        RLC A ;          7
1910        MOV P3.0,C
1911        SETB P3.1
1912        CLR P3.1
1913        RLC A ;          8
1914        MOV P3.0,C
1915        SETB P3.1
1916        CLR P3.1
1917
1918        SETB P3.6 ; DATA TO SHIFT REGISTER LATCH
1919        CLR P3.6 ; BLOCK SHIFT REGISTER LATCH
1920
1921        ; D TO A  MEMORY SELECTION
1922        MOV A,FUNC
1923        ANL A,#00001111B
1924        CJNE A,#00000101B,TXIT ; #00000101B = RIT
1925        MOV R1,#2DH ; #2DH EQUATES TO TUNHI
1926        LJMP DAIT
1927 TXIT:    ; AINT RIT, MUST BE XIT, #00001010B = XIT
1928        MOV R1,#2FH ; #2FH EQUATES TO DAHI
1929 DAIT:
1930        CLR P3.6 ; BLOCK PLL AND SHIFT REGISTER LATCHES
1931
1932        ; EXTERNAL INTERRUPT 1  D TO A
1933        CLR C
1934        MOV A,@R1 ; HIGH BYTE FIRST
1935        SWAP A
1936        ANL A,#11110000B
1937        CPL A
1938        CLR C
1939        RLC A ;          1
1940        MOV P3.0,C
1941        SETB P3.1
1942        CLR P3.1
1943        RLC A ;          2
1944        MOV P3.0,C
1945        SETB P3.1
1946        CLR P3.1
1947        RLC A ;          3
1948        MOV P3.0,C
1949        SETB P3.1
1950        CLR P3.1
1951        RLC A ;          4
1952        MOV P3.0,C
1953        SETB P3.1
1954        CLR P3.1
1955
1956        DEC R1
1957        CLR C
1958        MOV A,@R1 ; LOW BYTE FOR D TO A CONVERTER
1959
1960        CPL A
1961        RLC A ;          5
1962        MOV P3.0,C
1963        SETB P3.1
1964        CLR P3.1
1965        RLC A ;          6
1966        MOV P3.0,C
1967        SETB P3.1
1968        CLR P3.1
1969        RLC A ;          7
1970        MOV P3.0,C
1971        SETB P3.1
```

```
1972        CLR P3.1
1973        RLC A ;         8
1974        MOV P3.0,C
1975        SETB P3.1
1976        CLR P3.1
1977        RLC A ;         9
1978        MOV P3.0,C
1979        SETB P3.1
1980        CLR P3.1
1981        RLC A ;         10
1982        MOV P3.0,C
1983        SETB P3.1
1984        CLR P3.1
1985        RLC A ;         11
1986        MOV P3.0,C
1987        SETB P3.1
1988        CLR P3.1
1989        RLC A ;         12
1990        MOV P3.0,C
1991        SETB P3.1
1992        CLR P3.1
1993
1994        CLR P3.7 ; DATA TO D-A LATCH
1995        SETB P3.7 ; BLOCK D-A LATCH
1996        MOV READY,#30
1997
1998        POP 1H ; 1H = R1
1999        POP DPH
2000        POP DPL
2001        POP B
2002        POP ACC
2003        POP PSW
2004        RETI
2005
2006           END
2007
2008        ;......................................................
```

Addresses

The following firms may be useful for procuring parts for this project. I have experienced good service from the ones which I used. A few are included as alternate sources, but I have not had the opportunity to order materials from all of those listed.

Occasionally, some will want to know "who you are with." They are generally not impressed if you tell them "my kids and dogs." I've found that some sales personnel with wholesale-only firms receive commissions and are easier to deal with if you can buy through your employer, a friend's business, or some similar arrangement. Most firms listed here are happy to deal with individuals.

Allied Electronics
A Subsidiary of Hallmark Electronics
Corporation
This company uses branch offices.
A catalog can be requested from
(800) 433-5700.
Wide range of electronics parts including Motorola ICs

Amidon Associates, Inc.
P.O. Box 956
240 Briggs Ave.
Costa Mesa, CA 92626
(714) 850-4660
Fax (714) 850-1163
Ferrite and iron inductor cores

Barker & Williamson Corp.
10 Canal St
Bristol, PA 19007
(215) 788-5581
Fax (215) 788-9577
Inductors and filters

Circuit Specialist, Inc.
P.O. Box 3047
Scottsdale, AZ 85271-3047
(800) 528-1417
Assorted parts, microcontrollers, memory, computer hardware

Dan's Small Parts and Kits
Box 3634
Missoula, MT 59806-3634
(406) 258-2782
Assorted parts including variable capacitors

Digi-Key Corporation
701 Brooks Ave S
P.O. Box 677
Thief River Falls, MN 56701
(800) 344-4539
Fax (218) 681-3380
Wide range of electronics parts

Fox-Tango Corp: *See* International Radio and Computers, Inc.

International Crystal Mfg. Co.
P.O. Box 26330
10 North Lee
Oklahoma City, OK 73126-0330
(405) 236-3741
(800) 725-1426
Fax (800) 322-9426
Quartz crystal resonators

International Radio and Computers, Inc.
3804 South U.S. 1
Fort Pierce, FL 34982
(407) 489-0956
Fax (407) 464-6386

Jameco
1355 Shoreway Rd.
Belmont, CA 94002
(800) 831-4242
Fax (800) 237-6948
Assorted parts, microcontrollers,
memory, computer hardware

James Millen Electronics
P.O. Box 4215BV
Andover, MA 01810-4215
(508) 975-2711
Fax (508) 474-8949
RF inductors, capacitors

JAN Crystals
2341 Crystal Dr.
P.O. Box 60017
Fort Myers, FL 33906-6017
(800) 526-9825
(813) 936-2397
Fax (813) 936-3750
Quartz crystal resonators

JDR Microdevices
1850 South 10th Street
San Jose, CA 95112-4108
(800) 538-5000
Microcontrollers, memory, computer
hardware

Mini-Circuits Labs
P.O. Box 350166
Brooklyn, NY 11235-0003
(718) 934-4500
Mixers, amplifiers, signal combiners,
etc.

Mouser Electronics
2401 Hwy. 287 N
Mansfield, TX 76063
(800) 346-6873
Wide range of electronics parts

Needhams Electronics
4539 Orange Grove Avenue
Sacramento, CA 95841
(916) 924-8037
EPROM emulators and related
products

Newark Electronics
This company uses branch offices.
A catalog can be requested from
(800) 463-9275.
Wide range of electronics parts including Motorola frequency synthesizer
chips

Parallax Inc.
6359 Auburn Blvd.
Suite C
Citrus Heights, CA 95621
(916) 721-8217
EPROM emulators, microcontrollers,
and development systems

RADIOKIT
P.O. Box 973
Pelham, NH 03076
(603) 635-2235
Fax (603) 635-2943
Assorted parts

RF Parts Co.
435 S. Pacific St.
San Marcos, CA 92069
(619) 744-0900
Fax (619) 744-1943
RF semiconductors and related components

Sentry Mfg. Co.
P.O. Box 250
Chickasha, OK 73023
(405) 224-6780
(800) 252-6780
Quartz crystals

Universal Cross-Assemblers
9 Westminster Drive
Quispamsis, NB
Canada E2E2V4
(506) 849-8952
Fax (506) 847-0681
Microprocessor/microcontroller development hardware and software including meta cross-assemblers

624 KITS
171 Springlake Dr.
Spartanburg, SC 29302
(803) 573-6677
Kits and assorted radio parts

The following board fabricators are interested in producing bare (unstuffed) circuit boards for readers. They have a copy of the patterns, so it is not necessary to send a set. Please contact them for prices, availability, and shipping information.

Atlas Circuits Company
1500 Old Lake Road
P.O. Box 892
Lincolnton, NC 28092
(704) 735-3943

FAR Circuits
18N640 Field Ct.
Dundee, IL 60118
(847) 836-9148 voice and fax

First Proto
4201 University Drive
Suite 102
Durham, NC 27707
(919) 403-8243

Bibliography

Cote, Raymond GA. 1993. "The future belongs to steam." Robot Explorer 1(6):1–2.

DeSoto, Clinton B. 1936, 1981. Two Hundred Meters and Down: The Story of Amateur Radio. Newington, CT: American Radio Relay League, Inc.

Fifty Years of A.R.R.L., 1st edition. 1981. Newington, CT: American Radio Relay League, Inc.

Grebenkemper, J. 1993. "Ironing out your own printed-circuit boards." QST, July, 42–44.

Hayward, Wes. 1994. Radio Frequency Design, 1st ARRL edition. Newington, CT: American Radio Relay League, Inc.

Hayward, Wes, and Doug Demaw. 1986. Solid State Design for the Radio Amateur. Newington, CT: American Radio Relay League, Inc.

Intel Embedded Applications. EAN: 9781555122423. Available from McGraw-Hill.

Intel Embedded Microcontrollers. EAN: 9781555122300. Available from McGraw-Hill.

Motorola RF Data Manual Volume II, 5th edition. 1988. Phoenix, AZ: Technical Information Center.

RCA CMOS Integrated Circuits. 1983. Somerville, NJ: RCA Corporation.

Schetgen, Robert (editor). 1995. The ARRL Handbook for Radio Amateurs, 72d edition. Newington, CT: American Radio Relay League, Inc.

Shrader, Robert L. 1967. Electronic Communication, 2d edition. New York: McGraw-Hill.

Ward, A. 1987. "Monolithic microwave integrated circuits." QST, February, 23–29.

Index

Italics indicate illustrations.

347

Crunchy microphone syndrome, 5
Currents, division of, 96–97
Custom-ground crystals, 18, 19
Cutoff frequency, 7
  calculation of (formula), 7
CW signals (*see* Continuous-wave signals)
CW (*see* Continuous-wave telegraphy)
CW transmission:
  advantage over SSB signals, 293
  and Morse code telegraphy, 293–298

D

DAC (digital-to-analog converter) circuits, 93
  precise timing of, 96
  producing ideal waveforms with, 96
  use with DDS, 93
DAC output, eliminating glitches in, 97
DAC value, 231
Data, transference to display, 200
DDS (*see* Direct digital synthesis)
DDS system:
  calculation of, 98
  limitations of, 98–99
Debouncing routine, 213
Demodulation, 13
Detector/modulator, 1
Detector/balanced modulator, 14
Diagnostic display mode, 197–198
Dial cord, 28
Digital circuits, 87
Digital computer, control and coordination of circuits, 169
Digital electronic circuits, and signal processing, 114
Digital signal processing (DSP), 99
Digital spectrum analyzer, 44

Digital-to-analog (DAC) converter circuits, 93
Digital value, generation and translation of, 93–96
Diode mixer, 14, 18
Diodes:
  prevention of multiple switch closures, 195
  protective, 79–80
Diode transistor logic (DTL), 118
Dipmeter, 42
Direct conversion receiver:
  for on-air selectivity, 23
  using circuits as, 24, *25*
Direct digital synthesis (DDS), 87, 92–99
Direct register addressing, 212
Dirty signal, 264
Distortion, of waveform, 13
Double-sideband (DSB) transmission, 13–29
  simple phone transceiver, 1
  voice transmission and internal filtering, 26
Down-counter, 121
DRAM (dynamic RAM), 177
Drilling pattern for simple-sided copper-clad board, 11
Driver and power amplification, 64–67
DSB (*see* Double-sideband transmission)
DSP (*see* Digital signal processing)
Driver/final board, 64–67
Dual-bridge diode mixer, 14
DTL (*see* Diode transistor logic)

E

ECL (*see* Emitter-coupled logic)
80C31:
  arithmetic operations, 216–217
  binary numbers in operation of, 206
  components of, 178

80C31 (*Cont.*):
  Data transfer instructions, 218–221
  and 8-bit instruction cycle, 170
  external memory requirement, 175
  and frequency counting, 154–156
  hardware design and description of, 178–180
  internal use of binary system in, 175
  logical operations, 217–218
  and waveform generation, 200
80C31/8051 family:
  and boolean variable manipulations, 219–220
  instruction set, 210–214
Electret, 5
Electret microphone, 4, 5, 7
Electrical spark discharge, 88–89
Emitter-coupled logic (ECL), 130
EPROM, 216
  emulators, 234–236
  erasing, 234
Equate directive, 215, 224
EQUATE statement, 212
Erasable programmable read-only memory (*see* EPROM)
Exclusive-OR circuit, 199
External interrupt, 225
External oscillator, 181

F

Feedback, and phase-locked loop, 111–114
Feed-through capacitors, 246–247
Ferrite beads, 68
FET oscillators, 102
FETs, 31
  as storage elements, 178
  and variations in transconductance, 104
Field, 211
Field-effect transistors (*see* FETs)

349

Internal latch, and transister on/off, 180

Interpreted languages, 209

Interrupted light beams, in optical shaft encoder, 191

J

Jump destinations, 229

Jump instruction, 220–221, 225

K

Keyed carrier generator, 20–22

Kilobytes (KB), 177

L

Label field, 211

Labels, 229

LCD module, operation of, 197–198

LCDs, 199

LC (local oscillator) filters, 50–52

Least significant bit, 96

LEDs, 138–139, 190, 191, 194, 195

Light-emitting diodes (*see* LEDs)

Light paths, spacing of, 191–192

Liquid crystal displays (*see* LCDs)

LM383, 4

Local oscillator filters (*see* LC filters)

Logical decision-making circuits, 114–118

Logic circuits, 114–118

Logic families, 118

Logic gates, 115–123

building sequential logic elements with, 118–124

Loop filters, 124–127

and phase errors, 125–126

Loop gain, measurement of, 126–127

Lower sideband, 196

Low IF circuits, 31–45

testing the SSB generator, 40–42

construction and testing of, 39–42

and crystal filters, 37–38

matching and amplification of, 35–36

and receive mode signals, 31–32

and transmit mode signals, 33–34

and two-tone audio signal generation, 43–45

Low impedance winding, 67–70

Low intermediate frequency circuits (*see* Low IF circuits)

LSB (*see* Least significant bit)

M

Machine language, 206

Magnetic pump, 166

Main loop, major tasks of, 225–226

Mass-produced crystals, 19

Master-oscillator power-amplifier (*see* MOPA)

Memory devices, 177–178

Metal-oxide semiconductor field-effect transistors (*see* MOSFETs)

Microcontroller:

automating operations by, 87

contrast with microprocessor, 175

for managing operating-frequency parameters, 87

port pins and time-multiplexed communication, 187

and use of register banks, 212

Microphone, 4–8

all-salvage, 5–6

electret, 4, 5, 7

homemade, 5–6

ready-made, 5

Microprocessor-based controller (*see* Microcontroller)

Milliwatt operating, 63

Mirror-image pattern, 9

Mixer diode (*see* Diode mixer)

Mixers, 1, 14

balance and suppression of unwanted signals, 50

Mixer transformers:

winding, 15–16

Mixing, 13, 15

Mnemonics, 206–207, 211

Modulation, 13, 91–92

for communication, 91

MOPA, 89

Morse code telegraphy:

and mobile CW, 297

skill acquisition, 295–298

MOSFETs, 82

Motor controller, as capacitor shaft driver, 164

Motorola MC145158P circuit, 131–135

Multiconductor cable, 6

Multiplexer/demultiplexer, 76

N

Needham's Electronics, EPROM programmer, 233

Negative RF feedback, 65

Net, 22

NFET (N-channel field-effect transistors), 179

Nonferrous metals, and electromagnetic coupling, 243

Not-PSEN (program-store-enable pin), 176

O

Octal system, 170

Ohmmeter, 70, 194

use in funding bridge connections and shorts, 186

On-chip RAM, 177

Ones complement, 174

351

353

355

About the Author

Randy L. Henderson is an electronics technician with more than 20 years of experience in the design, fabrication, troubleshooting, and repair of many types of electric and electronic circuits. Mr. Henderson served as an assistant technical editor for the American Radio Relay League (ARRL) on several different book projects. He has also written many articles for popular electronics magazines such as *QST, 73 Amateur Radio Today, QEX,* and *Nuts and Volts*, as well as for the *ARRL Handbook for Radio Amateurs*.